STABILITY

OF

LARGE STRUCTURES

STABILITY
OF
LARGE STRUCTURES

K. A. Zalka

and

G. S. T. Armer

BUTTERWORTH
HEINEMANN

Butterworth-Heinemann Ltd
Linacre House, Jordan Hill, Oxford OX2 8DP

PART OF REED INTERNATIONAL BOOKS

OXFORD LONDON BOSTON
MUNICH NEW DELHI SINGAPURE SYDNEY
TOKYO TORONTO WELLINGTON

First published 1992

British Library Cataloguing in Publication Data
Zalka, K. A.
Stability of Large Structures
I. Title II. Armer, G. S. T.
624.17

ISBN 0 7506 0733 5

Library of Congress Cataloguing in Publication Data
Zalka, K. A.
Stability of large structures/K.A. Zalka and G.S.T. Armer
p. cm.
Includes bibliographical references and index.
ISBN 0 7506 0733 5
1. Structural stability. I. Armer, G. S. T. II. Title.
TA656.Z35 1992
624. 1 7--dc20 92-9293
CIP

Printed and bound in Great Britain

CONTENTS LIST

	Introduction (K A Zalka)	viii
	Notations (K A Zalka)	x
1	**Background** (G S T Armer)	1
1.1	Preamble	1
1.2	Historical development	2
2	**Characteristics of structural systems** (G S T Armer)	7
2.1	An analysis	7
2.2	The load-deformation characteristic	19
3	**Design for robustness** (G S T Armer)	23
3.1	Stability versus robustness	23
3.2	Current design methods	27
3.3	The prospects for current rules	31
4	**System stability** (G S T Armer)	33
4.1	Safety in systems	33
4.2	Robust systems	34
5	**Continuum method for regular frameworks under normal loading** (K A Zalka)	39
5.0	Introduction	39
5.1	Assumptions	41
5.2	Substitute column	42
5.3	Frameworks on fixed supports	44
	5.3.1 Frameworks subjected to concentrated forces at top floor level	44
	5.3.1.1 The continuum model	45
	5.3.1.2 Axial deformation of the columns	52
	5.3.1.3 Frameworks with variable cross-sectioned columns	55
	5.3.2 Frameworks with UDL at floor levels	58
	5.3.2.1 The continuum model	59
	5.3.2.2 Axial deformation of the columns	66
5.4	Frameworks on pinned supports	68
	5.4.1 The continuum model	69

5.4.2 Frameworks subjected to concentrated forces at top floor level 69
5.4.3 Frameworks with UDL at floor levels 72
5.5 Loading 80
5.5.1 Structures with concentrated loads and UDL 80
5.5.2 Correction for frameworks subjected to UDL 81
5.5.3 The effect of non-proportional loading 83
5.6 Accuracy analysis, characteristic deformations and design examples 84
5.6.1 Frameworks on fixed supports 85
5.6.2 Frameworks on pinned supports 92
5.6.2.1 Frameworks with ground floor beams 92
5.6.2.2 Frameworks without ground floor beams 97
5.6.3 Non-proportional loading 102
5.6.4 Test results 104
5.7 Conclusions 106

6 Generalization of the continuum method (K A Zalka) 109

6.1 Non-proportional frameworks 109
6.1.1 Frameworks with columns of different cross-section 109
6.1.2 Frameworks on pinned supports without ground floor beams 110
6.1.3 Frameworks with longer first storey columns 111
6.2 Coupled shear walls 113
6.2.1 Coupled shear walls subjected to concentrated forces 114
6.2.2 Coupled shear walls with UDL at floor levels 116
6.2.3 Design examples - accuracy analysis 117
6.2.3.1 Comparison with FE results 118
6.2.3.2 Test results 121
6.2.4 Conclusions 123
6.3 Cross-wall system buildings 124
6.3.1 Approximate solution 124
6.3.2 Design example 126
6.4 Analogy between frameworks and sandwich columns 128
6.4.1 Cantilever with shear and bending deformation 129
6.4.2 Approximate solution 130
6.4.3 Exact solution 133
6.4.4 Application to frameworks 135
6.4.5 Numerical example 138
6.5 Supplementary remarks 139
6.5.1 Semi-rigid joints 140
6.5.2 Elasto-plastic analysis 140
6.5.3 Generalization 141

7 Spatial analysis of regular structures under normal loading (K A Zalka) 143

7.1 Introduction 144
7.2 Substitute core 145
7.3 Uniformly distributed floor loads 149
7.3.1 The governing differential equations 150
7.3.2 Doubly symmetric layout 152
7.3.3 Singly symmetric layout 156
7.3.4 Unsymmetric layout 163
7.4 Concentrated load at roof level 167

7.5 Practical application 168
7.5.1 Formulae for the overall analysis of buildings 168
7.5.2 Design examples 169
7.5.2.1 System of support cores 169
7.5.2.2 A single support core 173
7.6 Supplementary remarks 178
7.6.1 Shear centre - centre of torsion 178
7.6.2 Pierced support elements 178
7.7 Conclusions 180

8 Appendices 183

8.1 Additive theorems (K A Zalka) 183
8.1.1 Southwell's theorem 183
8.1.2 Dunkerley's theorem 184
8.1.3 The Föppl-Papkovics theorem 184
8.1.4 The Rankine formula 185
8.2 Derivation of the governing equation for torsional buckling (K A Zalka) 185
8.3 Tables for torsional buckling (K A Zalka) 194
8.4 Mathematical modelling of test data (G S T Armer) 209
8.4.1 Preamble 209
8.4.2 Introduction 209
8.4.3 The elements of Law 209
8.4.4 Numerical proofs 212
8.4.4.1 Units 212
8.4.4.2 Operational techniques for measurement 214
8.4.4.3 Scale of measurement 215
8.4.4.4 The problem of dimensions 215
8.4.4.5 Constants 216
8.4.5 Law structure 217
8.4.5.1 Function 217
8.4.5.2 Law class 217
8.4.5.3 Class 1 Laws 218
8.4.5.4 Class 2 Laws 219
8.4.5.5 Class 3 Laws 219
8.4.5.6 Higher Class Laws 221
8.4.6 Discontinuities 221
8.4.7 The representation of empirical data by a simple law form 223

9 References (K A Zalka) 227

10 Bibliography (K A Zalka) 237

Author index (K A Zalka) 245

Subject index (K A Zalka) 249

INTRODUCTION

The aim of this book is to discuss complex phenomena in a simple way and to try to give answers to some questions in the theory of the stability of large structures. Where possible, simple solutions readily applicable in structural engineering design for stability problems are given. This is mostly the case when regular structures are subjected to normal loads. When the structures are non-regular or are under abnormal loads, no simple solutions are available. New aspects for the design philosophy and guidelines for practical structural design are presented for such cases.

It is also the authors' intention to draw attention to simple structural engineering methods which help to develop engineering common sense. This may seem to be a paradox in the age of computers when the solution of engineering problems is often believed to be only a question of time and the choice of the appropriate software package. This may not always be so since computer based methods and reality are often two different things. By the widespread application of computers, the gap is widening between practical structural engineers actually designing and supervising the building of structures and the scientist developing theories and procedures for the analyses.

This book does not claim to give a comprehensive state-of-the-art report on available theories and methods for the stability analysis of structures. Most of the ideas and methods presented are the results of the authors' own research. A wide range of papers, books and monographs is listed at the end of the book. The majority of them are direct references; others represent valuable sources to the background of structural stability and for further research.

The types of structures covered in the book are: Frameworks on different support systems, shear walls and coupled shear walls, symmetrical cross wall system buildings, buildings supported by a system of shear walls and cores of open or closed cross-sections and large panel system buildings.

The continuum method is advocated for solving the eigenvalue problems of the stability of regular structures. The different applications show that the method is reasonably simple and indicate that it can be applied to similar problems in other fields. Comparison is made between the available exact and different approximate solutions. Conventional additive theorems like the Southwell theorem and different forms of the Dunkerley theorem are used to establish basic characteristics of behaviour.

A number of design examples show how to use the formulae in practical applications. These numerical examples are worked out in detail, giving all the information needed for the actual structural engineering computation. The examples include frameworks, coupled shear walls, cross wall system buildings, column systems supported by a single core and support systems of shear walls and cores.

Background is given in Chapter 1, showing the development of the definition of stability and offering an alternative definition. Based on five propositions, the following Chapters give four general strategies for design for stability. These strategies are useful for every type of structure used in structural engineering and are essential for non-regular structures or for structures under abnormal loads where the conventional methods do not offer simple solutions for the structural engineer. Design for robustness is discussed in Chapters 3 and 4 where the important problem of progressive collapse is also explained. Theoretical and practical examples help one to understand how safety in complex systems can be achieved.

Chapter 5 deals with frameworks on fixed and pinned supports. After establishing the governing equations, closed formulae are presented for the critical load. These formulae make it possible to carry out the stability analysis in minutes. Not only is the procedure fast but it also shows the basic four factors which characterize the behaviour of the framework. The individual contribution of the characteristic structural elements to the overall behaviour of the structure being known, the way to the optimum structural system is given. As the formulae are approximate, a comprehensive accuracy analysis involving hundreds of frameworks of even extreme stiffness characteristics concludes this chapter.

Chapter 6 shows the different possibilities, which can be explored in applying the continuum method established in the previous Chapter for the stability analysis of regular frameworks to other problems. The possibilities include certain non-regular frameworks, frameworks with flexible beam-column joints, coupled shear wall systems, cross wall system buildings and elasto-plastic stability analysis. The most important of all is perhaps the application to coupled shear walls and cross wall system buildings. This is an area where, because the stiffness ratios are sometimes extreme, discrete methods often fail in the stability analysis. The analogy with sandwich columns is mainly of theoretical importance.

The complex problem of torsional-flexural buckling of regular structures is discussed in Chapter 7. The system of governing differential equations is set up and the different modes of buckling are established. Pure torsional buckling (the basic problem with spatial analysis) is discussed in detail and the importance of the torsional and warping stiffnesses together with other geometrical characteristics is shown. Closed form formulae are given for the critical load of pure torsional buckling. When the shear centre of the support system and the centroid of the applied load do not coincide, as is the case with many practical applications, pure torsional buckling and flexural buckling in both principal planes combine. The critical load is obtained from a simple formula for which the load factor is given in tables.

A summary of the conventional additive theorems (the Southwell, Föppl-Papkovics and Dunkerley theorems), the derivation of the governing differential equations of torsional-flexural buckling and ten tables for the load factor of torsional-flexural buckling are given in the Appendix. Aspects of the mathematical modelling of test data are also included.

NOTATIONS

for Chapters 5, 6 7 and 8

Capital letters

A	cross-sectional area, area of the ground plan of the building
A_b	cross-section of the beams
A_c	cross-section of the columns
$A_{k,ij}$	coefficients
A_l	local cross-sectional area
B	size of the ground plan of the building in direction y; horizontal displacement of the top of the infinitely stiff column on pinned supports
C	centroid; constant
E	modulus of elasticity
E_b	modulus of elasticity of the beams
E_c	modulus of elasticity of the columns
F	applied concentrated load
F_{cr}	critical concentrated load
F_{cr}^{bu}	critical concentrated load for a building
F_{cr}^{e}	elastic critical concentrated load
F_{cr}^{ep}	elasto-plastic critical concentrated load
F_{cr}^{p}	rigid-plastic critical concentrated load
F_{cr}^{*}	critical load of a one-storey high column
$F_{cr,x}$	Euler's critical load in plane xz
$\bar{F}_{cr,x}$	critical concentrated load in plane xz, taking shear deformation into account
$F_{cr,y}$	Euler's critical load in plane yz

$\bar{F}_{cr,y}$ critical concentrated load in plane *yz*, taking shear deformation into account

$F_{cr,\varphi}$ critical concentrated load for pure torsional buckling

$\bar{F}_{cr,\varphi}$ critical concentrated load for pure torsional buckling, taking shear deformation into account

$F^{t}_{cr,\varphi}$ critical load for pure torsional buckling, associated with the torsional stiffness

$F^{\omega}_{cr,\varphi}$ critical load for pure torsional buckling, associated with the warping stiffness

F_g global bending critical concentrated load

F^*_g global bending critical concentrated load for one-bay frameworks

F_l local bending critical concentrated load

$\bar{F}_l$ local bending critical concentrated load for frames with variable cross-sections

G modulus of elasticity in shear

H height of building

I second moment of area

I_b second moment of area of the beams

I^*_b second moment of area of the beams for coupled shear walls

I_c second moment of area of the columns

I^u_c second moment of area of the columns at the top

I^b_c second moment of area of the columns at the bottom

I_g global second moment of area of the columns

I_l local second moment of area of the columns

I_o polar moment of inertia

I_x second moment of area with respect to axis *x*

I_y second moment of area with respect to axis *y*

I_{xy} product of inertia

I_ω warping constant

J St. Venant torsion constant

K shear stiffness of frameworks, shear critical load

K^* shear stiffness of coupled shear walls, shear critical load

K_b stiffness of the beams; global shear stiffness, global shear critical load

$\bar{K}_c$ stiffness of the columns; local shear stiffness, local shear critical load

L	width of the framework; size of the building in direction x
M	bending moment
M_t	torque
N	total applied uniformly distributed load (UDL), measured at ground floor level
N_{cr}	critical UDL
N^b_{cr}	critical UDL in bending
N^{bu}_{cr}	critical UDL for a building
N^D_{cr}	critical UDL with Dunkerley correction
N^s_{cr}	critical UDL in shear
$N_{cr,j}$	critical UDL of the j-th support element
$N_{cr,x}$	critical UDL in plane xz
$\bar{N}_{cr,x}$	critical UDL in plane xz, taking shear deformation into account
$N_{cr,y}$	critical UDL in plane yz
$\bar{N}_{cr,y}$	critical UDL in plane yz, taking shear deformation into account
$N_{cr,\varphi}$	critical UDL for pure torsional buckling
$\bar{N}_{cr,\varphi}$	critical UDL for pure torsional buckling, taking shear deformation into account
$N^t_{cr,\varphi}$	critical load for pure torsional buckling, associated with the torsional stiffness
$N^{\omega}_{cr,\varphi}$	critical load for pure torsional buckling, associated with the warping stiffness
N_g	global bending critical UDL
N_l	local bending critical UDL for frameworks on fixed supports
N^*_l	fictitious local bending critical UDL for frameworks on pinned supports
O	shear centre
Q	intensity of the UDL on the floors [kN/m^2]
S	stiffness; shear stiffness of sandwich columns
S_i	member stiffness
S_1	size of ground plan of cross wall system building in direction x
S_2	size of ground plan of cross wall system building in direction y
T	shear force

Small letters

a multiplier
b multiplier; size of support core
b_k coefficient in a power series
c depth of beam in coupled shear walls; multiplier
c_i constant of integration
c_k coefficient in a power series
d multiplier; location of fictitious zero stiffness
e number of storeys
g parameter for columns with variable cross-sections; global axis
h storey height
h_o effective length of one-storey high columns
h^* effective length of the longer first-storey columns
$\bar{h}$ irregular first storey height
i parameter running from 0 to n
i_p radius of gyration
k parameter running from 0 to ∞
k_b distributed stiffness of the beams
$\bar{k}_b$ stiffness of the beams with semi-rigid joints
k_c distributed stiffness of the columns
k_j joint stiffness
l frame-span=distance between the axes of columns; distance between walls; local axis
m reduction factor for semi-rigid joints; section border; uniformly distributed bending moment; number of support walls and frames in the building
m_z torque
n number of columns; number of walls; number of cores; number of terms in a power series; section border
p intensity of UDL on the beams
q intensity of UDL along the columns
q_{cr} critical intensity of UDL along the columns
r reduction factor
s width of walls; bay in a cross wall system building
t distance between the shear centre and the centroid of the vertical load; wall thickness; global centroidal axis of the cross-sections of the framework
t_i distance between the *i*-th column axis and the centroidal axis of the columns
t' reduced wall thickness of a core
u translation in direction x
v translation in direction y
z vertical coordinate axis; coordinate
x horizontal coordinate axis; coordinate

$\bar{x}$ horizontal coordinate axis; coordinate in coordinate system $\overline{xy}$

x_o coordinate of the shear centre

y horizontal coordinate axis; coordinate

$\bar{y}$ horizontal coordinate axis; coordinate in coordinate system $\overline{xy}$

y_o coordinate of the shear centre

y' first differential quotient

Greek letters

α eigenvalue, critical load parameter for frameworks; critical load parameter for pure torsional buckling

α_S Southwell's critical load parameter

β stiffness ratio for frameworks; stiffness ratio for pure torsional buckling

$\bar{\beta}$ stiffness ratio for cantilevers with bending and shear stiffnesses

γ stiffness ratio for torsional-flexural buckling

δ stiffness ratio for torsional-flexural buckling

Δ error of the approximate procedures

ε critical load parameter for monosymmetric cross-sections, eigenvalue,

λ critical load parameter for columns with bending and shear stiffnesses; critical load parameter for unsymmetric cross-sections, eigenvalue,

λ_F the Föppl-Papkovics critical load parameter

ν multiplier; the ratio of the critical load to the applied loads

ρ constant depending on the shape of the cross-section

σ compressive stress

τ non-dimensional parameter for monosymmetric cross-sections

τ_x non-dimensional parameter for unsymmetric cross-sections

τ_y non-dimensional parameter for unsymmetric cross-sections

φ rotation

1

BACKGROUND

1.1 PREAMBLE

A definition of engineering in construction was given by Brunel in his investigations of ship-building practice -

> "no materials shall be employed on any part except at the place and in the direction and in the proportion to which it is required and can be usefully applied for the strength of the ship ..."

This has been one of the ideals striven for by engineers since Man began to exercise control over his environment. Two special problems faced the early builders. Firstly, were the materials available to them adequately strong to perform the tasks required and secondly, were the structures they created for shelter, bridging and defence, sufficiently stable not to fail in use or to cause any hazard to life. It is perhaps an error to imply that "material strength" was identified as a problem, since it seems more likely that "element strength" would have been seen as the critical factor. For example, the use of a tree trunk to bridge over a stream or as part of a cruck for a house generates no need for any detailed knowledge of the strength of the timber, although experience with differing species would soon lead to the use of preferred types [Granger, 1934]. It is reasonable to deduce that the early steps in structural design would have been sustained by a qualitative understanding of materials and of structural form. In this context, the concept of structural form would have been strictly limited to considerations of element shape and to the difficulties and possibilities of connection between the elements. This, in modern terms, apparently grossly circumscribed database sustained construction through the Egyptian, Greek, and Roman periods of our history and on to the magnificent era of mediaeval church construction. Not until the new science of the 16th and 17th centuries was there any possibility for a significant or substantial change in the philosophy on which structural design could be based.

The new science established the supremacy of the philosophical device of reductionism as a tool to analyse the natural environment. The new ideas provided a basis for the development of simple modelling of phenomena which could be observed by everybody, or more realistically, by anybody with a will to see the World in a new light. In this context, the term simple is related only to the form of the models and not to the concepts which underlay them. It seems amazing to those of us involved in modern research, that any development of natural philosophy could be sustained without careful observation of nature. Particularly as the information forthcoming from such a source is essential for the foundation of a model or law to describe (and to predict) the world about us. Nevertheless, there have been periods in our history when our most influential thinkers have considered that the simple observation of what was going on about them was irrelevant for the business of philosophy and that empiricism was an inappropriate and unnecessary activity.

1.2 HISTORICAL DEVELOPMENT

So, those concerned with the development of structural design were eventually presented with some new tools - the motivation to make careful observation of existing structures and structural elements and of materials suitable for new construction, and the analytical methods appropriate for the interpretation of the resulting information. As with all other theories, principles tailored for one application are not wholly satisfactory when appropriated for another, even if there appear to be a host of common characteristics which should sustain such a common usage.

Leaving aside the undoubtedly important theological incentives for the creation of the new science [Descartes, 1968], it was also established to provide an understanding of the world in which Man found himself. Natural phenomena could be observed directly. Naturally occurring materials and objects could be put into controlled environments and their characteristics, as exhibited in these specific situations, recorded for future analysis. The scenario was entirely that of an outside observer looking in through a window and analysing what he could see. In many ways therefore, the new methods were unsuitable for structural design which, as its title suggests, is a **synthetic** process. At best, the strengths of materials and of some structural elements are accessible to an analytical science, but what of the millennia of experience of structural form which had sustained the process of design prior to these developments? It must have taken some brave men to reject such a successful technique.

The designer is required to create a structure which will satisfy his clients demands for functional performance and aesthetic appeal. Thus the critical element of design, structural form, that is an essential part of the design process and which calls upon the engineer's creative skills, cannot sensibly be accommodated by the classical tools of science. Of course, many clever and valuable devices such as the use of iterative analyses, have been developed to paper over the cracks left by the "new design method" but it is never possible to do the really impossible. It is for this reason, that engineers accept that there is still "art" as well as "craft" in their apparently pragmatic profession.

We have therefore the somewhat paradoxical situation that, prior to the introduction to designers of the scientific method, they were able to build what can only be described as some of the great wonders of the world without any detailed scientific knowledge of the properties of their construction materials. Likewise, following the introduction of this method, yet more magnificent structures were constructed, - this time without any scientifically based theories of structural form. Unfortunately, these facts say more for the engineers than they do for the theories. Obviously, this is a somewhat black and white picture of the situation. Nevertheless, it is evident that some soundly-based principles related to the form of a structure would be an asset to the designer.

At the beginning of this Chapter, strength and stability were identified as the essential elements of design from the point of view of the structural engineer. On the one hand, there is a vast store of information on the strengths and other characteristics of materials suitable for application in construction. To a lesser extent, there is information on the performance of common structural elements i.e. principally real or apparent strengths. There is also a very restricted fund of information on the properties of real and complete buildings subject to a limited range of environmental conditions. On the other hand, for the second critical feature of design, that is stability, we are faced with the rather strange situation. In spite of the hundreds of books and papers written on the subject, there is hardly a useful qualitative definition of what is meant by the word stability and no quantitative one at all. A recent paper [Komarakul-na-nakorn, 1990] lists some mathematical definitions but these are not really accessible to most designers and are illustrative of the fact that mathematical models are directed at single point conditions rather that the general characteristics of a system. Such approaches are appropriate for the analyst but only an indirect aid to the designer.

Perhaps the most widely accepted interpretation of the term stability is in the context of stable equilibrium and unstable equilibrium. Consider the case of a stick-like element, such an element laid between two blocks to form a bridge may be called stable whereas, if it is balanced on one end on a flat surface, it will be unstable (Fig. 1-1). From this simple model comes the most universally accepted definition of stability, viz:

> A system is said to be stable provided that a small agitation does not cause large disturbances of the system.

This particular form of words was used by Renton in his discussion on the stability of frames [Chilver, 1967]. He further refines it for mathematical interpretation, by saying that a system is stable where "finite changes in it are not produced by infinitesimal perturbations". It is arguable that neither definition is of great help to the engineer at his drawing board or even in front of his computer terminal, who is trying to produce a "stable structure" since there is no route to a quantitative solution to the problem. A more recent definition is [ISE, 1988]:-

Provided, that displacements induced by normal loads are acceptable, then a building may be said to be stable if:
* a minor change in its form, condition, normal loading or equipment would not cause partial or complete collapse.
and
* it is not unduly sensitive to change resulting from accidental or other actions.

The first element can be described as little more than a good description of the concept of stability as perceived by its authors. It seems that the twenty years of research and development work which separated the invention of the two definitions, which incidentally included the notorious failure at Ronan Point in London's East End [Griffiths, 1968], did not appear to produce a more practical and usable improvement. The second element however incorporates a quite new dimension by making reference to accidental actions. This development brings together the traditional concept of stability and the new concept of robustness. It is perhaps appropriate to link with this concept, the even newer term "structural integrity."

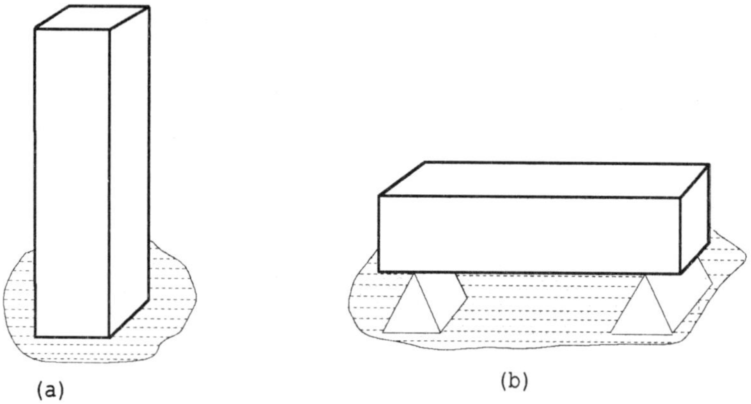

Fig. 1-1. Demonstration of simple stability concepts

There is a clue to the nature of the problem of the traditional definition in the simple model shown in Fig. 1-1. The change in the condition of stability of the element is related exclusively to the support conditions and to its orientation in space. This could be interpreted as its use either as a beam, in which case it is considered to be stable, or it is used as a column, and is considered to be unstable. It is not difficult to develop this model into one with several elements which can be arranged and connected in a variety of ways. Some of the arrangements will be stable and some will collapse with the slightest disturbance. Therefore it seems quite reasonable to make a connection between the arrangement of the elements and the resultant stability. Accepting that since the elements themselves do not fail, there is also no demonstrable relationship between material or element strength and the system failure. Inevitably, one is lead to the conclusion that there is a direct correlation between structural form and stability. This deduction is not particularly original and as a statement of principle, somewhat raw and of limited use in this form. It is of special interest though, since it creates a positive link between the tools and explicit knowledge of the modern designer with the understanding of his equally successful (in their own terms) predecessors.

The principle relating structural form to stability is firmly embedded in several of the major codes of practice. It is usually invoked by implication with such phrases as "use a good planform"; "provide continuous ties horizontally and vertically throughout the building"; "put expansion joints at a spacing of not less than 30 metres" and so on. Such guidance is always either in a prescriptive form or is exhortatory. These design rules are essentially founded on an empirical based understanding of the behaviour of structures. Nevertheless, they are quite appropriate for common structural forms but of much more restricted applicability for unusual structures.

It is worth exposing briefly the use of the term **stability**. In current usage, there is considerable ambiguity between the uses of the term made by designers in different contexts. Firstly, there is the traditional situation related to critical conditions of equilibrium, for example the possibilities of overturning of structures acting as monoliths, for example dams. Secondly, the situations related to critical changes of load-carrying mechanism such as exhibited in Euler buckling. Thirdly, the potential risk of progressive collapse in complex structures subject to local accidental damage and is related to "robustness." Finally, the problem of structural integrity often of concern in existing structures where there is a need to interpret actual structural details and make some assessment of likely performance in the future. Of these four uses, the first is fairly straightforward to deal with and is not considered further here. The second has been the subject of numerous studies mostly concerned with the performance of elements or simple structural frames. This area is developed for more complicated structures in Chapters 5, 6 and 7. The third area identified i.e. robustness is the subject of the next two chapters and is not easily linked to the previous applications. The fourth usage is currently normally used for existing structures and is not dealt with in any detail here.

2

CHARACTERISTICS OF STRUCTURAL SYSTEMS

2.1 AN ANALYSIS

Before the use of the scientific method as a basis of the design process for structures, concepts of ideal proportion and 'the golden section' [Morgan, 1961] inter alia were used. Particular shapes such as circles, and equilateral triangles were assumed to be endowed with desirable characteristics both aesthetic and others and were used in a variety of combinations to control the proportions of a building (Fig. 2-1, Project for a church by Philibert de l'Orme [Mainstone, 1959]). At first sight, such notions seem rather outlandish to the modern engineer. However, the characteristics of shape and proportion can be related to the behaviour of a structure. Consider for example a structure in the form of a monolithic element, then using the idea of Euler buckling, it can be seen that a short fat structure failing under a compression load will be much more stable than a tall thin one (of similar construction) under the same load condition. There are it seems then many clues which point towards a significant and direct connection between what is generally understood by the word "stability" and the shape and form of a structure or structural element.

In spite of the widespread success of form-controlled design (Fig. 2-2), the application of such abstract and rather metaphysical inventions that found a place in the work of the early builders, somehow seems out of place in the field of contemporary construction design. In order to progress, it is useful to analyse the problem in simple steps and to use simplistic models of structures to illustrate each stage of the discussion. Consider the stability of any structure comprised of an assembly of prescribed constituent elements, there are two pertinent questions which can be posed on the basis of such limited information. Firstly, does the arrangement of the elements affect its stability? Secondly, is there a limit to the number of elements arranged in a particular manner which is naturally stable? Figure 2-3 shows the results of some simple experiments. Figure 2-3/a is a wall-like structure built with model clay bricks in stretcher bond and fourteen courses high. This model was subjected to a disturbance which caused a collapse leaving just three courses intact (Fig. 2-3/b). Figures 2-3/c and 2-3/e illustrate two tower structures, the first with a triangular arrangement of bricks 34 courses high and the second with a square arrangement of bricks and 51 courses high. Both these models were also subjected to a disturbance resulting in a stable unit of 14 courses for the first and 22 for the second. Thus here is a simple demonstration that the arrangement of elements may be directly correlated with the stability of a construction, and further that for a particular arrangement, there is a critical maximum number of elements for a stable structure. It is probable that the precise number of elements in a stable construction will also depend on the size and nature of the disturbance causing a failure. The models therefore highlight a critical feature of the problem of stability, namely, that stability is a characteristic of a **structure-environment** system. In this context, environment includes all the loads and load combinations.

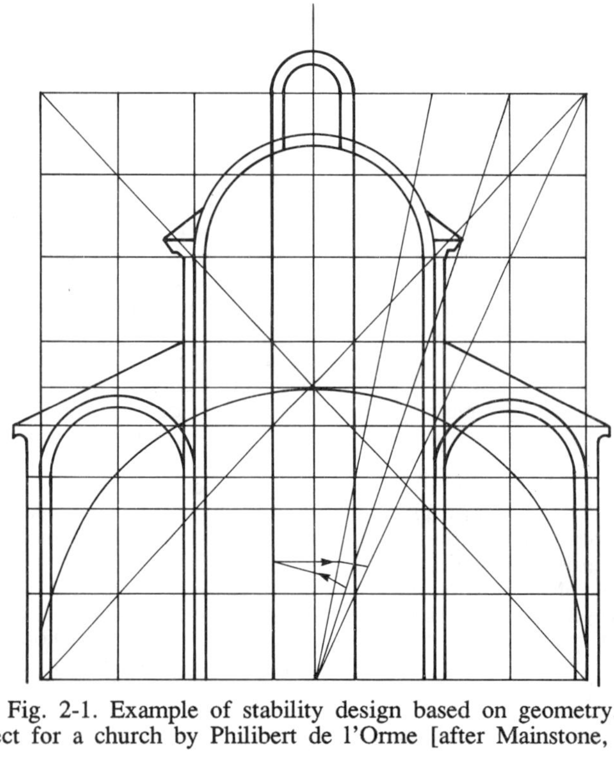

Fig. 2-1. Example of stability design based on geometry.
Project for a church by Philibert de l'Orme [after Mainstone, 1959]

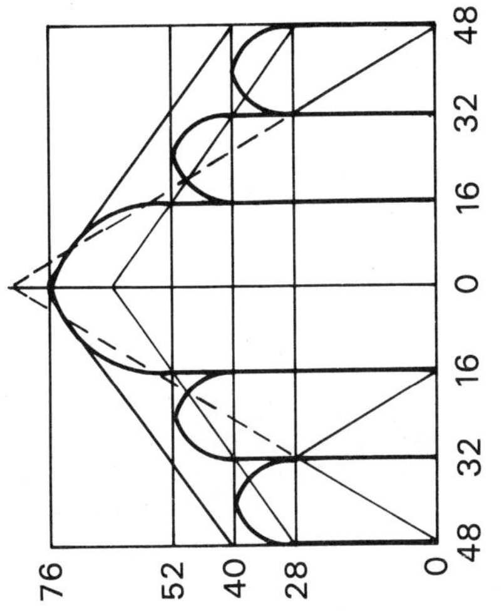

Alternative proposals for the cross section of Milan cathedral. Top left, project of 1891 'ad triangulam' with heights as calculated by the mathematician Stornoloco. Bottom left, project of 1392 'ad quadratum.' Top right, accepted project of 1392 'ad figuram triangulam' above the lower part of the 1391 project.

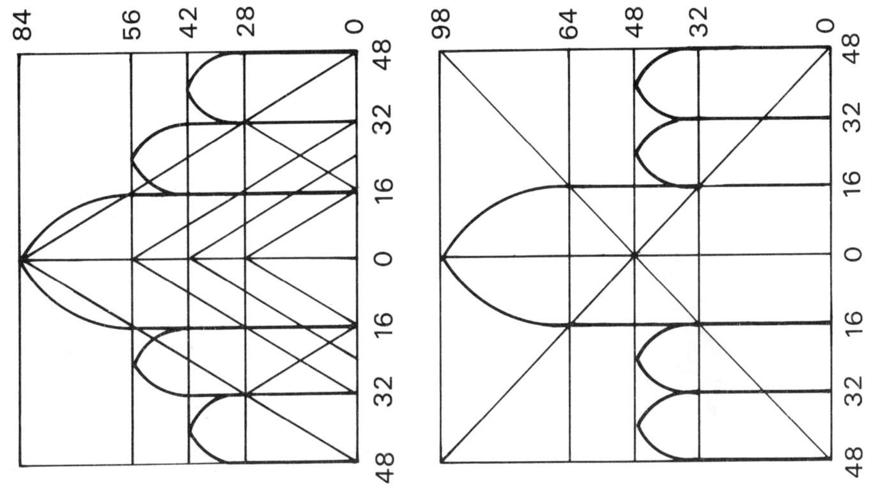

Fig. 2-2. Preliminary designs for Milan Cathedral

Fig. 2-3. Demonstration of the effect of the arrangements of units on the size of a stable structure. - Continued next page

e f

Fig. 2-3. Demonstration of the effect of the arrangements of units on the size of a stable structure

Another more realistic set of models is shown in Fig. 2-4. In these cases, plain frames have been constructed in a common form with identical square bays and with similar weights of material in each model. Only the distribution of material is varied for each frame. The variations are as follows for the six models:

1) The beams and columns have identical sections
2) Two bay by five storey heavy sections
3) Four heavy columns with similar sections for the other columns and all the beams
4) Two bay by five storey heavy sections with weak central core

All these models were tested under vertical loading and the variety of response is clear to see. Model 1 failed at one storey level, and since all the other storeys were to all intents and purposes identical, then although the test was stopped, the model exhibited an unstable or brittle failure mode. Model 3 with the strong vertical columns at two bay centres and weak linking beams, failed as a complete entity. Such a mode of failure involving all the material has the most stable or ductile characteristic of all the models tested.

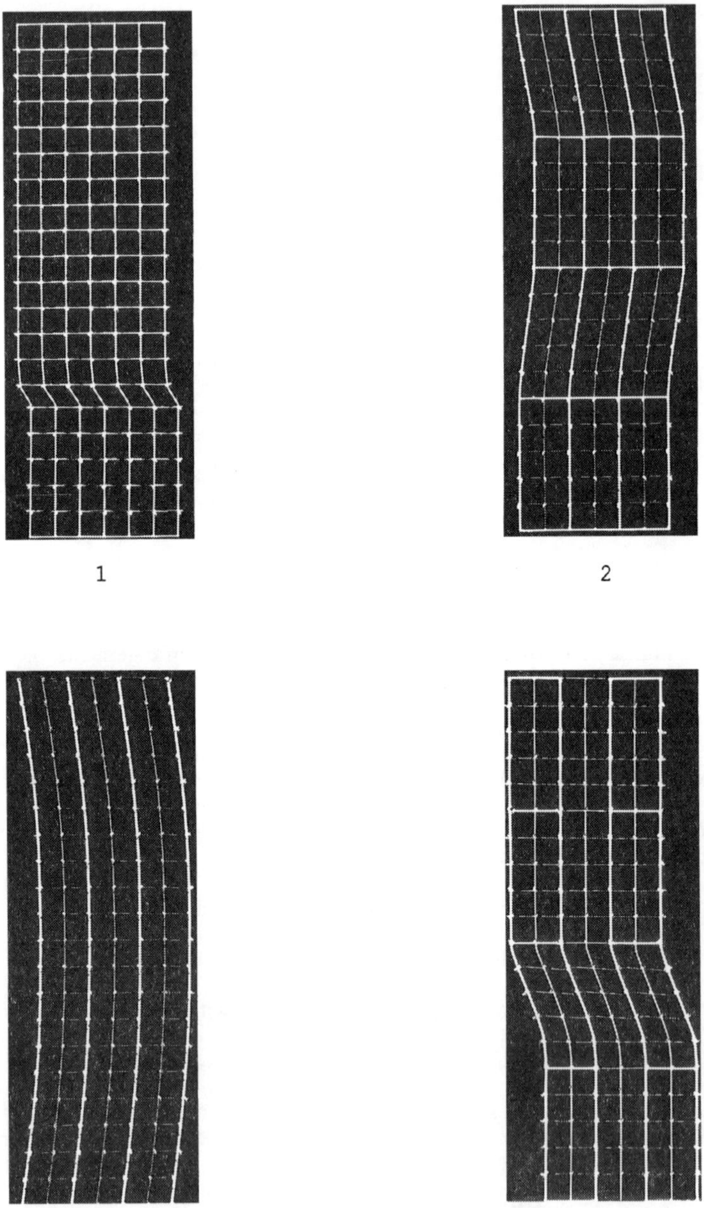

1 2 3 4

Fig. 2-4. Stability tests on model plane frames under vertical loading

It is interesting to observe in these models that there is also an illustration that elements collected into assemblages/structures in an arbitrary way with the constructor taking no heed of the individual properties of the elements, function in groups with an obviously common response to environmental change. This phenomenon is best seen in Models 2 and 4. These model tests suggest that there is an important similitude between the concept of ductility and that of stability, which will be called upon later.

Another interesting aspect of the models is the failure loads reached during testing. The table below gives the values of the peak loads weighted according to the amount of material in each frame - which, for practical reasons, varied slightly for each arrangement.

Model 1:	0.113
Model 2:	0.064
Model 3:	0.059
Model 4:	0.048

As might be expected, Model 1 which exhibited the most brittle behaviour, was also the strongest.

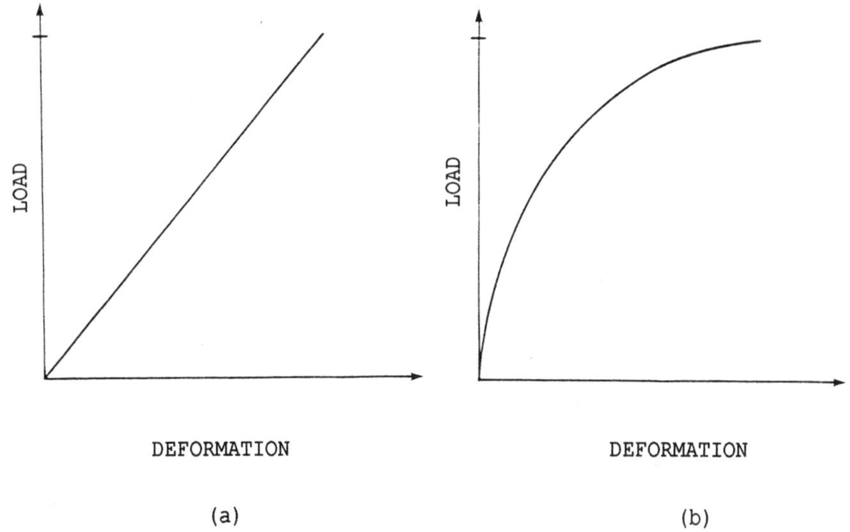

Fig. 2-5. Schematic representations of load-deformation characteristics

Moving for the moment to another direction of attack on the problem, consider the load-deformation characteristic of a structure as represented by a graph. Tests on structures or structural elements are usually simple in the sense that they comprise measurement of response to single load types, such as point loads on flat slabs or uniformly distributed loads on beams or floor planks. The results of such tests may be modelled usefully by continuous single line graphs. Such graphs are usually regarded as reflections of particular absolute characteristics of the structures which generated them.

Figures 2-5/a and 2-5/b show two typical forms of such graphs. The first shows a characteristic which would be commonly ascribed as belonging to a structure which exhibited instability. The second would not, and interestingly would not necessarily be described as ductile, except in those countries where earthquakes are a significant problem and structural ductility has been identified as a legitimate defence against this particular form of loading. It is now useful to look at these two curves more closely to see what sort of behaviour they reflect. The first is simply a straight line to a maximum of both the abscissae and the ordinate. The second is a monotonically increasing curve with a single maximum value. Now treating these graphs as independent characteristics, what can be said about the system which generated them? From such limited information two important properties can be deduced.

a) There is a finite upper limit to both systems.
b) For a change in input *dl* the corresponding change in *dd* is **independent** of the load level for Fig. 2-5/a and is **dependent** on the load level for Fig. 2-5/b.

The second property (b) can be re-interpreted as the load-carrying mechanism for system *5a* is constant throughout whereas, for system *5b* the load-carrying mechanism changes. Now if it is accepted that Fig. 2-5/a does indeed represent a system which suffers from instability, then it seems reasonable to suggest that *5b* represents a stable system. This being so, then it can be seen that in order to achieve stability in a system, the facility of multiple load-carrying mechanisms must be available.

Before discussing deformation characteristics further, it is useful to reconsider some commonly held ideas pertaining to the behaviour of structures under load. Firstly, take the relationship between the properties of a structure and the parallel properties of the materials from which it is made. The design engineer is usually concerned, inter alia, with the density, strength, and elastic properties of his structural materials. Taking as an example density, it is clear that there can be no general one-to-one correspondence between the density of a material and that of a structure made from that material. Whilst the mass of the structural material usually provides the mass of the structure, its form will be the most significant parameter in the determination of its density. In general there will be an order of magnitude difference between the two values, since the whole point of building in the first place is to provide space! Similar parallel arguments can be adduced to identify the weak relationship between material strength and elasticity and the corresponding properties of the whole structure, since structural form is also the major factor in the determination of these characteristics of complete buildings. Secondly, another important property of buildings which is of special interest to the designer is its ductility. It is usually assumed that ductility or brittleness of the construction material or at least of the elements is an essential requirement to produce similar characteristics in the structure. Consider however, the case where virtually none of the material is involved in a failure, as might be the situation for a structure with simple joints (a pack-of-cards house would be the **reductio ad absurdum** in this argument), then there could be only an insignificant correlation between material and structural properties. Alternatively, in cases where all or most of the material in a structure is involved in its failure, then a one-to-one relationship between material and structural properties could exist. Together, these arguments lead to the following proposition:-

Proposition 1

> It is neither sufficient nor necessary to specify the ductility of a structural material in order to control the stability of a construction made of that material

Structural engineers put most of their design effort for the assessment of the tripartite system of load/structure/foundation into the potential behaviour of the structure. This part is that over which most control can be exercised, both in terms of the connection with the other parts of the system and also within the structure itself. The earlier discussion of structural stability and ductility suggests that a general relationship between independent load-paths and stability exists. It follows then that the independent redundancies in a structure which provide alternative paths for load transmission dictate the degree of ductility that it will exhibit. The number of potential load-paths within a structure is related to the number of structural elements in the building, to the number of potential load-carrying mechanisms within the elements and the joints between them, e.g. bending, shear and torsion etc. and to the arrangement of the elements. Any particular loading system will only mobilise a subset of the paths available. For a structure to behave in a ductile/stable manner, there must be a high degree of independence between each path and a trend towards greater strength as each new mechanism comes into play. High connectivity within these paths increases the probability of retaining the load within the system. We now have the case for a further proposition:-

Proposition 2

> The stability of a structure-environment system depends upon the redundancy within that system

An essential feature of the phenomena of structural instability is movement. Movement of a structure during failure is obvious, for example, column buckling, and rotations at plastic hinges. More often than not, the structural displacements and effects associated with instability can be observed by the untrained eye. The same is not necessarily true for the complementary movements of the loads. Four categories of load movement can be identified easily as follows:-

a) Physical movement of loads from one part of the structure to another, for example, water ponding on flat slabs or debris loading on damaged structures, as in Ronan Point [Griffiths, 1968].

b) Transmission of loads or actions through media external the critical section of the structure. In practice, this medium is often another part of the structure, as in the case of a mining subsidence or of the Camden school roof [BRE, 1973]. In this latter example, the relatively strong transverse connections between a number of precast concrete roof beams of the school (Fig. 2-6) invoked a mechanism such that when one beam failed its load was transferred to a neighbour, and so on, causing failure of the whole roof structure. Bridging in granular and powdered materials in silos and in stacks of bricks or blocks are examples of load redistribution through the loading material itself.

Fig. 2-6. Roof failure at Camden School

c) Change of direction of action relative to the structure, for example, a switch from membrane to bending action during small deflections of a slab, or a switch from axial to buckling/bending behaviour in sway frames.

d) The imposition of new loads. These might be vehicle impact, explosions, other accidental loads or even additional loads of the types already present in a structure.

These movements lead to a third proposition:-

Proposition 3

> The failure of a structure requires the movement of loads or reactions within the structure-environment system

These arguments can be developed now into a discussion of those aspects of the complete structure/environment system which can accommodate change. It is change within a part of a system which is necessary to precipitate instability. Conversely, it is change within the complementary part of that system which will be sufficient or not to maintain the stability of the whole. The mechanism of this facility can be illustrated by the case of a uniformly loaded, simply supported beam.

Whilst the beam supports its design load it is stable. If the load is then increased up to the ultimate capacity of the beam, such a change will be accommodated by increased deformation and response. The system can only remain stable above this load level if some major change occurs such as the provision of a prop, the direct strengthening of the beam or a switch from bending to catenary action. Obviously, the timing of such changes is critical for them to be effective, thus there is a need for feedback of information to ensure that the extra strength is provided before the load increase takes place. Likewise, should the strength of the beam be reduced as a result of a degradation process, then suitable information feedback procedures will be required to implement a maintenance programme to ensure continued stability. The same principle is used in other parts of construction. For example, design codes and standards establish bounds which allow variations in the system up to prescribed levels. Inspections, legal waivers, or other devices are available to facilitate desirable or necessary variations to these bounds and to sustain stability within the population of structures. This proposition follows from the discussion:-

Proposition 4

> The stability of a structure-environment system depends upon two essential characteristics:-
>
> 1. The potential capacity of each part of the system to sustain change without failure.
> 2. The potential of the complementary part of the system to sustain change including the loss of a failed part.

These four propositions allow the development of some general strategies for design for stability.

1) **The provision of multiple, independent load paths or redundancy within the structure.** With most common forms of load, this strategy will directly affect the ductility of a structure. Structures comprising frames with infill panels are examples of designs following this approach. Many supposedly non-loadbearing elements will, in practice, contribute to alternative load-carrying mechanisms.

Fig. 2-7. The first stage of failure of the Stepney swimming pool roof

2) **The inclusion of devices to allow the building to avoid carrying load.** Examples of this approach include load-shedding elements such as windows for venting explosions, weak joints in walls and floors to prevent the horizontal transmission of actions e.g. expansion joints, anistropic construction. The latter is illustrated by the Stepney swimming pool roof [Bate, 1974] where weak transverse connections between adjacent precast concrete roof beams allowed single beams to fail without overloading neighbouring beams and causing the collapse of the whole roof (Fig. 2-7).

3) **The provision of local or global increases in resistance to enhance the overall strength of the structure.** This strategy is manifested by the tying clauses in many modern codes of practice, which seek to increase the overall strength of a structure by the use of improved connections between components. This device will not necessarily alter the brittle characteristic of a structure nor the nature of its final failure. However, by increasing the size of the trigger load needed to precipitate failure in the event of an accidental overload, it will reduce the likelihood of instability.

4) **The installation of environmental and performance monitoring and control systems.** These can ensure that a building is used as the designer intended by rapid feedback of information to prevent misuse, and that repair and maintenance is instigated following non-critical damage, and that unplanned for loads are avoided.

The application of these strategies must be judged according to each particular problem. One or more of them must be adopted on the grounds of economics and pragmatism. Some judgement has always to be exercised when attempting to establish the active and the potential load-carrying mechanisms in a designed structure.

2.2 THE LOAD-DEFORMATION CHARACTERISTIC

In order to develop a design methodology on the basis of the analysis in the previous Section, it is useful to interpret the ISE [1988] definition of stability and for the moment call it an interim definition since the final objective must be a fully quantifiable statement. A qualitative definition is:-

A structural system is stable if it has multiple mechanisms and/or paths which function sequentially, to carry the imposed loads and reactions safely.

It is worth noting that the essence of this definition differs in substance from the concept of robustness, which is often used when resistance to accidental loads is being discussed. A dictionary definition of robust is "strongly built". Obviously, structures have to be strong enough to carry their imposed loads, however, stability is related to the manner in which they do so. A further critical feature of the definition, is that it is potentially quantifiable. Figure 2-8 illustrates some load-deformation characteristics and shows the direction of increasing stability. Readers familiar with mechanical testing will notice that their experience would lead to an expectation of all three lines being combined to form a typical experimental line such as that shown in Fig. 2-8. This general form of characteristic is quite important and will be discussed later.

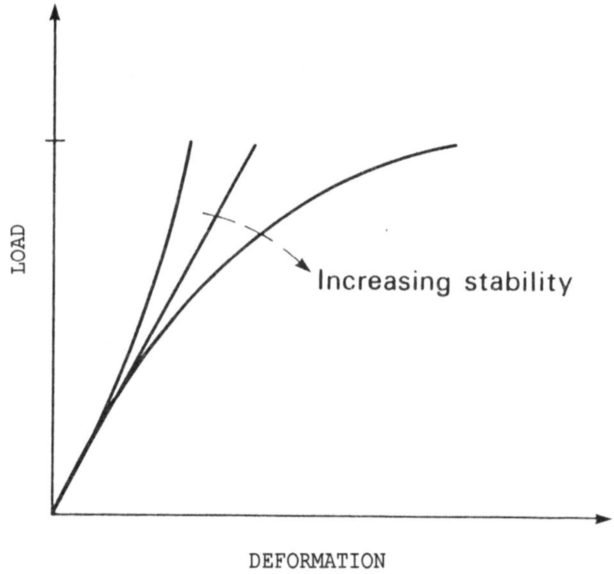

Fig. 2-8. The range of simple load-deformation characteristics

The two outside graphs in Fig. 2-8, that is, the one curving towards the *L*-axis and that curving towards the *D*-axis, both represent situations where the size of the structural response is dependent upon the existing load level. The first of these can be compared to a strain-stiffening material characteristic and the second to a strain-softening one. Both satisfy the definition of system stability, in that they reflect the employment of different load paths or load-carrying mechanisms which depend upon the load level. However, the choice of increasing stability from left to right of the figure, that is from load-stiffening through elastic to load-softening, corresponds to the universal search for ductility in situations where stability, in the conventional sense, is deemed to be critical. An alternative description is that our intuitive perception of stability in complex structures requires a significant feedback of information from an overloaded structure to the environment (e.g. the occupier). Clearly in these terms, the load-stiffening characteristic is most undesirable since it represents a situation where the deformational feedback is greatest for a high reserve capacity of load and least for a low reserve capacity, that is immediately prior to failure. Consequently, the corresponding response from the complementary part of the system is the smallest near the most critical region.

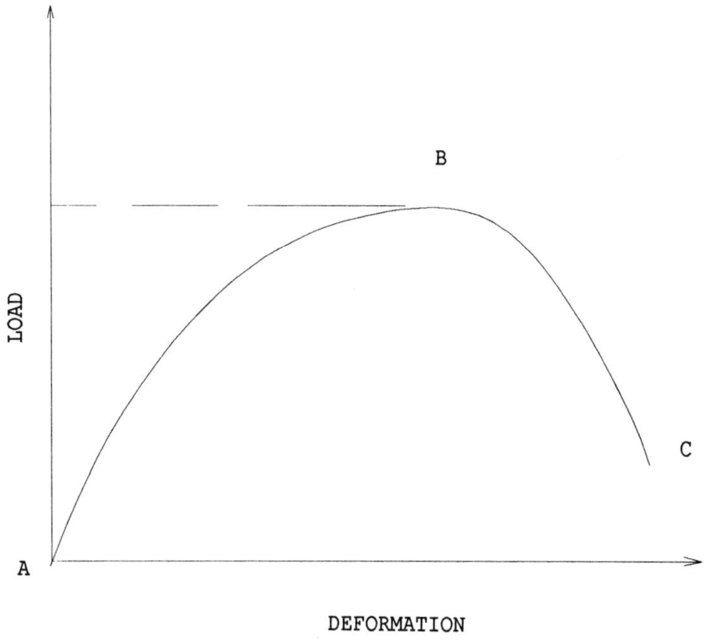

Fig. 2-9. Typical general form of experimental load-deformation characteristic

This general discussion on the shape of the graph generated by plotting values of an applied load against the response of a structure has avoided one frequently used element of such plots. Figure 2-9 shows a commonly illustrated form of graph. The part from *A* to *B* has been considered already, whereas the part from *B* to *C* has not, and it is this part which has an especial significance in the overall behaviour of the load-environment system. The acquisition of a part *B-C* in the laboratory is of special interest. For such a purpose, it is necessary to have a strain-controlled machine for load application, since the slope of the graph here is dependent upon controlling the rate of energy release into the test specimen. The sources of this energy are manifold. Firstly, the load application system, that is the hydraulic pump, gravity loads etc. Secondly, the elastic energy stored in the testing machine, reaction frame etc. Thirdly, the elastic energy stored in the undamaged parts of the test specimen. The section *B-C* represents the performance of the structure during the failure stage when its ability to carry its maximum load has been lost, and control of the load-specimen response has moved from the specimen to the environment of the failing zone, hence the requirement for a strain controlled machine in order to follow the degraded response curve. It is clear then that the gradient of the part *B-C* will depend on the response time of the loading machine and on the elastic nature of the environment surrounding the failing zone, and that

it **should not** be used as a characteristic of the specimen alone [Armer, 1989].
These various considerations lead to the following proposition:-

Proposition 5

> For monotonically increasing applied loads, only the section of the load-deformation characteristic from zero load to the maximum may be regarded as being representative of the properties of the loaded structure.

This is a most important point for those involved in the design of structures to resist earthquakes, since the temptation for them to invoke the apparent post-peak properties of the construction materials is the greatest and so are the potential risks of failure.

3

DESIGN FOR ROBUSTNESS

3.1 STABILITY VERSUS ROBUSTNESS

It is perhaps clear from the preceding chapters, that there is a conflict of understanding related to the meanings attached to the word "stability". On the one hand, there are the engineers and users of buildings who would like their structures to perform satisfactorily with ductile characteristics, under a variety of loads and load combinations. On the other hand, there are the mathematicians and researchers with a theoretical bent who have, possibly perforce, restricted their activities to a study of the properties of structural systems at singular instants and for single loads or load cases. Stability is the term used by both camps to identify the particular characteristic of the systems they are dealing with and thus there is ambiguity.

The difference between the two approaches is well illustrated by the quotation by Hunt [Thompson and Hunt, 1982] from a Russian paper [Andronov and Pontriagin, 1937] who identified a requirement that for phenomena in systems to be observable, they should be:-

"**robust** against **small** perturbations."

This is a nice and compact definition for the theoretician's concept of stability. In contradistinction, the structural designer's objective can be identified as the need to create building systems which are:-

"**robust** against **large** perturbations."

The structural engineering concept of "robustness" has developed over the last twenty years, although good designers will have had an intuitive appreciation of the need for their structures to meet the requirements for robust design. As with all ideas, it is always possible to find historical intimations of understanding. There can be no doubt however, that the failure of Ronan Point [Griffiths, 1968] triggered the most comprehensive attack on the problem as evidenced by the selected references given at the end of this chapter.

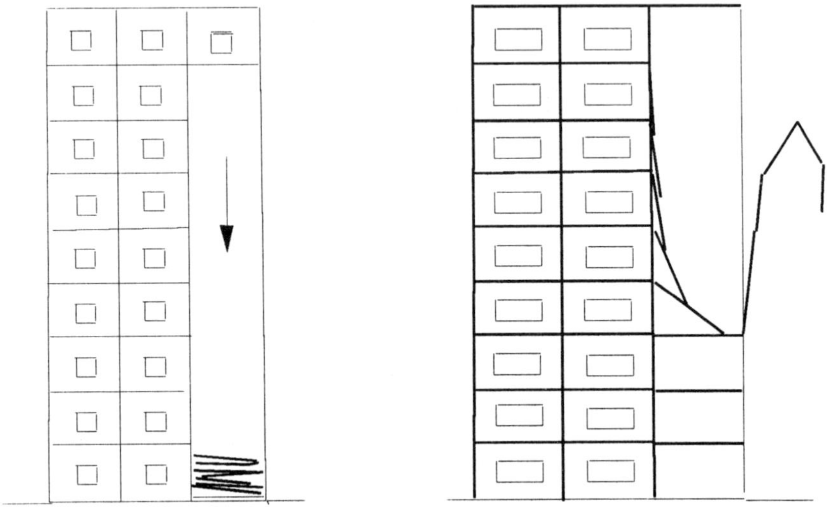

Fig. 3-1. Failure modes for high level local damage site

In the aftermath of the accident, the failure mechanism now named "progressive collapse" was considered to be of urgent practical concern for designers. This mechanism was deemed to occur when a structural failure seemed to be disproportionate to its original cause. It is implicit in this definition that an unidentified source of energy has been activated which is sufficient to cause significant structural damage. The only energy resource of adequate magnitude which is available in the system is the gravitational force mass of the building and its contents. In more practical terminology, it can be seen that the process of progressive collapse involves the conversion of parts of a structure into debris loading on other undamaged areas which in their turn add to the load impinging on the next section and so on. Two simple observations follow from this description. It is generally more likely that failure will progress from accidents near the top of susceptible structures (Fig. 3-1) since in such situations debris loads automatically fall onto the next floor. If, however, an identical accident occurs near the bottom of the building, the debris falls to the ground and is not available to create a progression of failure up the building. This lesson was expensively learned by a contractor who when called in to demolish a twenty storey redundant housing block, distributed his explosives around the second, fourth, sixth and eighth floors. On ignition, he found that as the building sank, the debris heap grew and the whole collapse process ground to a halt (Fig. 3-2). He was then left with a twelve storey building sitting on a large heap of rubble at a slight angle! Whilst the situation had some comic strip humour for those not involved, undoubtedly most people would take it as axiomatic that causing damage at the base of a building would create the most serious risk of wholesale collapse.

Fig. 3-2. The effect of local damage near the base of the structure

It is possible to have progressive failure moving up buildings (Fig. 3-3), but it clearly would be a relatively stable dynamic action and very susceptible to stopping if a slightly stronger-than-average section is met by the failure front.

Horizontal progressive failures are not often discussed. They are however, not unusual. Two examples are of particular interest. In 1973 the Skyline Plaza development was under construction in Fairfax County, Virginia [Leyendecker and Fattal, 1977]. The centre piece was to be two 23 storey towers of apartments separated by a four storey carpark. The construction was in-situ reinforced concrete for the towers and in-situ prestressed flat plates for the garage. The towers and the garage were under construction concurrently. By the time the construction gang was up to the 16th floor of one of the tower blocks it had become so efficient that the planned sequence of propping the floor under construction and the three below, did not allow enough time for the concrete to gain sufficient strength before the props began to be removed. The consequence of this was a failure of the structure around the crane area and a progressive failure to the ground. At this stage of construction, the first two storeys of the garage had been built - each storey was about 300 m x 250 m on plan. Around ten per cent of the top slab received debris damage from the tower failure, this created very large loads on the columns periphery of the damaged zone which then punched through leaving the slab load to be carried by the next set of columns which also punched through, and so the failure progressed across the deck with increasing loads on each new set of columns at the advancing failure front. The end result is shown in Fig. 3-4. Both the constructed decks failed since the top one having failed dropped onto the one below.

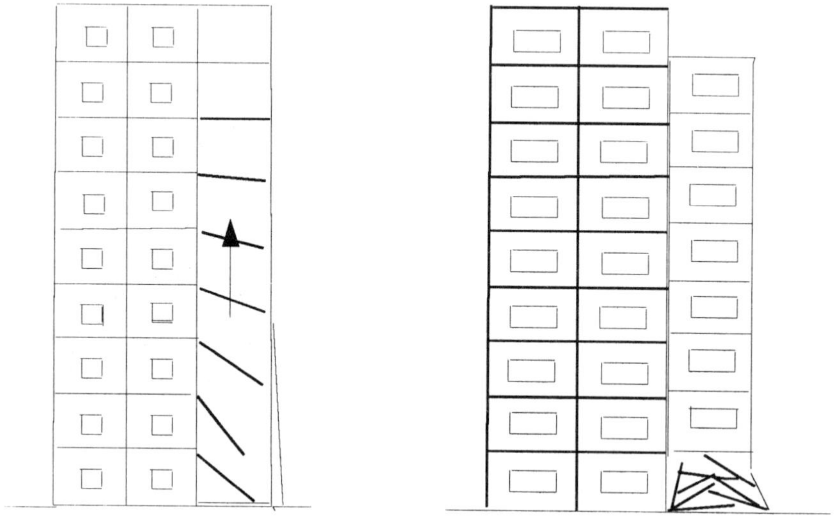

Fig. 3-3. Failure modes for low level local damage site

The second example of horizontal progressive failure occurred in Zurich in Switzerland [NCE, 1985]. The building involved was a public swimming pool. The roof structure of the building supported a suspended ceiling which was hung on rather complicated multi-element and multi-metal fixings. The combination of the environmental conditions in the roof space and the mixed material design of the fixing created ideal conditions for corrosion problems. These duly occurred and once the failures started they progressed until the whole ceiling had collapsed. Two particular points of interest can be drawn from these examples. Firstly, both the prestressed concrete plate of the garage deck and the ceiling element of the swimming pool had a high level of horizontal continuity. Once the support for parts of these plates had failed their loads were transferred to the next supports and so on. It can be seen then that the only way to prevent this mode of failure is to break the horizontal continuity by incorporating appropriate joints in the construction.

Fig. 3-4. Horizontal progressive failure

3.2 CURRENT DESIGN METHODS

Unfortunately for the writers of codes and other advisory documents, when pressed into dealing with this newly perceived problem there was no extant design philosophy on which to base design methods for engineers to use. The vast majority of the new design ideas were pragmatic (quite properly so) and empirical. Although it has to be said, that the experimental validation of proposed new design rules was generally limited to the assessment of local performance rather than their effects on the overall performance of multi-storeyed structures.

Not surprisingly, the development of the new rules was predicated on the adoption of a domestic gas explosion as a design parameter. One of the first strategies investigated was therefore, the use of pressure venting through building elements such as windows [Mainstone, 1971 and 1973]. This never proved to be a very convincing strategy on which to base design rules for housing structures constructed of precast concrete large panels, particularly since Ronan Point had large non-structural panels in each room which had patently failed to provide adequate pressure relief. The other equally significant problem associated with providing potential vents, is the need to provide prescribed maximum strengths for the venting elements and their fixings. This requirement is opposite to that for other load cases and brings extra conflict into the decision-making process.

The consensus opinion has arrived at three prescriptive design rule classes. The first calls for horizontal and vertical ties to be provided throughout the building. In the major British design codes the general requirements are stated thus:-

BS 8110: Part 1 1985. Structural Reinforced Concrete [BSI, 1985b]

> Unreasonable susceptibility to the effects of accidents may generally be prevented if the following precautions are taken.
>
> a) ...
>
> b) All buildings are provided with effective horizontal ties:
> 1) around the periphery;
> 2) internally
> 3) to columns and walls
>
> c) ...
>
> d)...This is generally achieved by the provision of vertical ties...

BS 5950: Part 1 1990 Structural use of Steelwork in Building [BSI, 1990]

> Every building frame should be effectively tied together at each principal floor and roof level.
>
> All column splices should be capable of resisting a tensile force of not less than two thirds of the factored vertical load...

BS 5628: Part 1 1985 Structural use of Masonry [BSI, 1985a]

> ...The general recommendations are aimed at the limitation of accidental damage and preservation of structural integrity...
>
> ...provision within the structure of vertical or horizontal tying, or both, is recommended.

In each case, the general clauses are supported by detailed advice regarding sizing and fixing of the ties. The second class of design rule comprises those rules which accept the notional removal of individual structural members. The designer is then required to re-analyse the building to ensure that the loads previously carried by the removed element can be carried by the adjacent structure. These techniques are the so-called "alternative path" methods. Associated with them is a legal requirement limiting the acceptable amount of consequential damage following the loss of the member. In the U.K., this is currently 15% of the floor area of a storey or 70 m^2, whichever is the least. The third class includes rules for the direct design of structural elements to resist a prescribed static pressure. The current value for this is fixed at 34 kN/m^2 (5 psi). The method by which this figure was derived is rather interesting. In the aftermath of the Ronan Point accident, a number of tests were undertaken on items such as biscuit tins and fuse box covers to simulate the damage caused to similar items by the explosion. There were also some theoretical analyses related to other damage. This work provided a range of static equivalent pressures from 12 psi to 3 psi which might reasonably be taken to represent the actual dynamic pressures experienced during the accident. With this data in hand, the authorities decided that 5 psi (34 kN/m^2) was a practical design value and it has remained at that level ever since.

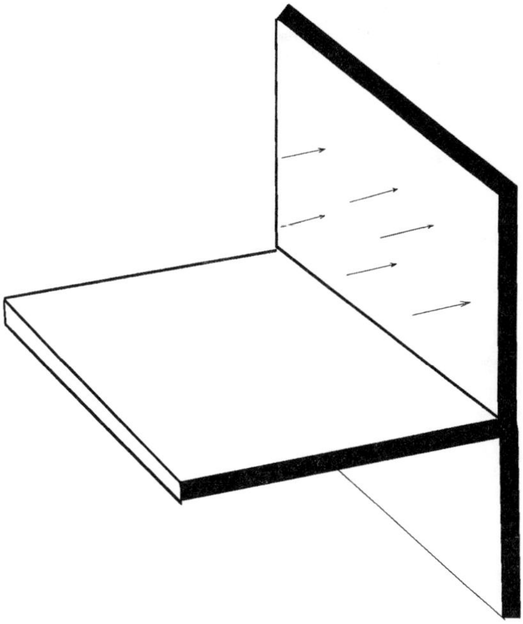

Fig. 3-5. The development of tie forces normal to crosswalls

When these rules were first introduced, direct design using the prescribed pressure was a rare occurrence. Designers found it relatively easy to provide ties or if this was not so, to follow occasionally, then to follow the alternative load-path route. More recent developments in construction practice involving the common use of much larger spans has however led to greater difficulty in meeting the prescriptive rules in classes one and two.

It is interesting to consider the possible effects of these rules on the actual behaviour of structures designed according to their recommendations. The general strategies for the design of stable systems developed in Chapter 2 provide a useful background against which to assess their efficacy. Considering the tying rules first, it is worth noting that the recommended design values for horizontal ties were originally derived from two calculations. One was derived from the forces needed to restrain a panel fixed at ceiling and floor level and loaded by a pressure of 5 psi (Fig. 3-5). The second was derived assuming the loss of a single load-bearing precast concrete panel and the forces required to allow the next panel above the lost one to cantilever over the space (Fig. 3-6). The dimensions of these panels were based predictably on those of the Ronan Point system. So the values for the tie forces to be used in new construction have come from an amalgam of the principles of the rule classes two and three.

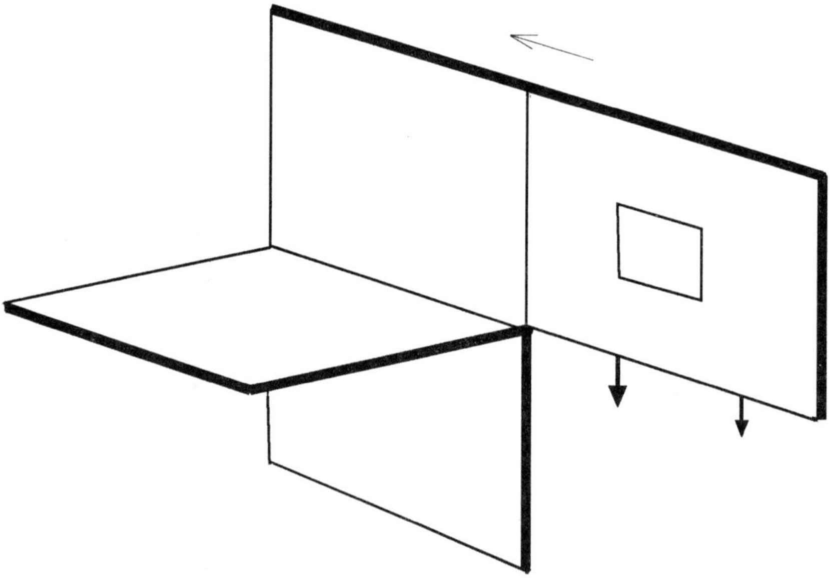

Fig. 3-6. The development of tie forces parallel to crosswalls

It seems reasonable to deduce that the principle effect of the application of the tying rules to provide an overall increase in the resistance of the building to accidental damage (Chapter 2 Strategy 3). This is obviously a legitimate approach up to the stage where actual damage is caused. Prior to such a failure the ties control the resistance of the structure to the accidental load. Subsequently, and before the ties can aid the development of alternative paths for the in-situ structural loads, they provide unwanted alternative paths for the accidental loads. For large accidental loads it is therefore quite possible that they will exacerbate the situation by enabling the development of failure zones significantly greater than might have occurred in their absence.

Although the argument, on which the alternative path rules are based, assumes that local damage has occurred, the actual effect on design is to introduce more connections and sometimes more structural material. The consequence for overall structural behaviour will therefore necessarily be similar to that produced by the tying rules.

The direct treatment of accidental loading by the specification of a load related to a notional situation provides strengthened elements and connections. It carries with it the implicit assumption that the designer could not reasonably be expected to prepare for any greater actual loads which might occur. Thus if the latter situation did arise, it could only be described as "unfortunate!" It is rather difficult to generalise about the effect of these direct design rules on the overall behaviour in the event of a local failure. The reasons for this difficulty stem from the fact that the designer, when applying them, is required to identify "key" or "protected elements" for the special treatment and will often use the other rules on other parts of the structure. One thing that is clear is that all three rule classes will usually lead to the provision of extra connections and often, extra structural material. Both these additions to the structure will increase the load-carrying mechanism potential of the structure. This fits the Strategy 1 developed in Section 2.1. Unfortunately, in this context, their effects are indeterminate.

3.3 THE PROSPECTS FOR CURRENT RULES

What is the future for the existing prescriptive rules to control the robustness of a structure? It is arguable that accidental or abnormal loads are very rare phenomena. It is equally arguable that large **complex** structures built outside earthquake zones and hurricane-prone areas will inevitably fail from some initial local defect. There is then really no doubt that design for robustness must remain in the structural engineer's vocabulary. Equally, there is a growing body of information which signals a need to re-assess the current rules. There are two major problems to be addressed.

Firstly, is a domestic gas explosion an adequate model for all the probable causes of local damage in large structures? The answer must be no when the potential range of causes is identified. The principle hazards are fire, degradation, impact and explosion. It is probable that this list is in the right order to reflect the relative importance of each category. This monograph is not the right medium to promote new prescriptive design rules, however, even if it were, the research on the response of complex structures to fire and explosions is still in hand and will not be available for practical application for some years yet and to change design rules without very sound new data is exceedingly counter-productive not to say potentially dangerous.

Secondly, the prescribed limits on acceptable damage following the loss of an element or part of a structure seem to be inappropriate for much construction now. If these limits are removed, the justification for accepting apparently lower standards of safety has to be found, also, the means by which key or protected elements are to be identified must be established. It is suggested that many of the elements currently having to be designed as key elements are in buildings unlikely to be subjected to explosions of domestic gas and that they are most often located at or near the ground floor level. If these two reasons are sustainable for particular situations then there are at least good grounds for not treating the elements as key elements.

4

SYSTEM STABILITY

4.1 SAFETY IN SYSTEMS

The word "systems" is used because the requirement for the occupants of a building to be safe and satisfied can be met by controlling the building environment as well as the physical structure of the building itself.

The concept of safety in current structural design is extremely complex in comparison with that which held sway in the days of simple factors on loads or material strengths. The danger or risks inherent in structures arise, in part, from the designer's limited knowledge of the exact environment in which they will function, the properties of the materials used in construction, and manner in which the loads will be carried. In the context of building design and construction, safety is sought by using the most appropriate design and construction techniques, the best materials and and the best environmental data. Within the design calculations, safety margins are ensured more specifically by the use of numerical factors.

The development of analytical theories based on elastic material properties and later, on plastic material properties inevitably lead to the ideas of limit state design. Directing the designer to consider his proposed structure at load conditions associated with "normal use" i.e. serviceability, and at collapse (or the ultimate limit state) inevitably results in ambiguity. The problem with multi-targeted actions as shown by René Thom [1975] is that it is only possible to satisfy single prescribed objectives. The mathematical models of natural phenomena lead to apparent "catastrophes", in real life, designs are controlled by only one of the limit states. Although the limit state philosophy places equal importance upon meeting serviceability conditions as upon meeting ultimate conditions, it is arguable that the principal objective of design should be to produce serviceable structures and that ipso facto, they would be safe. What then is the rôle of ultimate loads and load factors? It could be said of course, that the requirement for stability under extreme load states provides a specific security margin over the serviceability condition for the explicitly considered load cases. The evaluation of the margin does not, within the conventional meaning of such a factor, result in a partial safety factor. That part of the assumed safety control mechanism is more akin to design requirements in that it will affect primarily, structural behaviour.

The explicit use of numerical factors in the design process is to circumvent the differences between the design model and its assumptions, and the real structure and its environment. Factors in this class reflect ignorance of the real world in which the structure will have to survive. Consider the ideal "perfectly safe" building, such a building would be one in which the material and structural properties were known exactly and could be maintained throughout its life, and also for which the environmental, loading and support conditions where known and controlled within prescribed levels. No safety factors would be required for such a building!

Traditionally, designers have incorporated factors into their analyses by assuming pessimistic values for their basic variables, by the use of conservative analytical models and quasi-global factors at appropriate stages of a design. In such an approach, it is only possible to decide qualitatively whether or not the method is successful, by a study of the rate of failure in the population of buildings designed according to its principles. There is no practical way to evaluate the actual factors for any given load case.

The most recent device pressed on the designer, with a view to enabling him to control the safety level of his structure, is the result of the application of probability theory to the elements of the design process. The case for this introduction has been founded on the basis that in the real world, there will be random variations of many elements comprising the structure-environment system, and that the statistical procedures developed into the theory of probability enable the designer to deal with these variations in a rational manner.

The ground rules for the success of this development are :-

1. There should be available a large body of well-chosen data related to each basic variable it is required to factor.
2. There are clear unequivocal rules for the quantification of individual factors.

Unfortunately, there is very little data for most variables; the notable exception being wind loads. It also seems that the rules for quantification of the factors have not been well developed [Armer et al, 1987]. This has resulted in lengthy discussions on values for partial factors being the ultimate cause of much of the difficult negotiations related to the preparation of the Eurocodes.

4.2 ROBUST SYSTEMS

It has been suggested in Chapter 3 that one of the principal goals of the modern designer is to create structural systems which are robust against large perturbations. It is implicit then that in a qualitative sense, the concept of stability for an element is similar to robustness for the system. From an obvious argument, it can be said that what is a large perturbation for an element is just a small perturbation for a system - it is all a matter of appropriate scales - hence the rationality of using "stability" for both element and system.

The strategic requirements for robust systems were set out in Chapter 2. The first of these stated the need for multiple, independent load-paths or redundancy within the system. Redundancy is normally discussed in the context of framed structures. It should be remembered that in real structures, although individual elements may be designed as beams or columns etc., they all will be able to function in torsion, shear, compression, tension and bending. Whether or not they are able to exhibit all their potential will depend upon geometry and joints. A classic example of the potential capacity of an element to adapt its response to its condition, is the two-way spanning reinforced concrete slab. If such a slab is simply-supported around its edges, and uniformly loaded, the applied load will initially be transmitted to the supports by compressive membrane action which involves a central zone in compression surrounded by a ring in tension over the supports. The second stage which comes into play as the load increases is the normal bending action on which the designer will have based his analysis. The third stage will be tensile membrane action which has a central zone in tension surrounded by a ring of compression over the supports. This is a prime illustration of the statement that "structures will do everything they can to stop falling down." A statement that not only is true but also somehow contradicts the ethic of the analytically based design process.

For particular structures, the redundancy within individual elements is to a certain extent controlled by the capacities of the joints between neighbouring elements. To this extent, the joints themselves become elements. Then, of course, the number of neighbouring elements is important, since the more connections an element has, the more potential for acting in different load-carrying rôles it has. An interesting exercise to illustrate the effect of this "connectivity" in a structural system is to consider a large number of beam/column elements, say, two thousand erected in two three-dimensional frames of one thousand elements each - one tall slender structure and one short squat one. It is reasonable to postulate that the slender will be much more prone to instability failure than the short one. Now, the joints at the corners will have four elements meeting, whilst the joints on the faces will have five, and the internal ones, six. It can be seen that simply multiplying the numbers of each joint-type by the appropriate number of elements it services i.e. 4, 5 or 6, and adding these three numbers for each structure, will give a lower connectivity for the slender structure than for the squat one. In real structures, it is not easy to identify those redundancies which can be mobilised in the event of an abnormal load occurring.

This brief introduction to structural connectivity suggests that an appeal to general systems theory would be the most practical way forward - and so it may be in the future. There are, however, some considerable problems with its use in the analysis of structural systems now. The most successful practical application of the theory is in the field of data transmission through networks. This is essentially involved with moving data packets. To convert loads/stresses into a similar single class of data equivalent is a) extremely difficult and b) quite different from the familiar ground of the structural designer. It therefore does not seem to be a promising way forward at the moment for the practising designer, although the research community may find it a rich field.

An alternative approach is the classical method. That is, treat building structures as "material" specimens and classify them according to their brittleness, since brittleness, instability and lack of robustness seem to relate to a common characteristic. As suggested in Chapter 2, system robustness or stability could be represented by the changing curvature of the load deformation characteristic. There has been a long history of using power curves to model causal relationships. This use has been particularly successful in dimensional analysis which in turn has found its greatest application in the field of fluid mechanics. It has been of less value in the field of structural engineering [Wood, 1955]. The mathematical argument to sustain the universal applicability of the power curve form has been clearly established by Bridgman [1931] and Ellis [1966]). However, neither of these authors tackled the problem of how to gather and treat empirical data to ensure that it was in the right form to meet their assumptions. In Section 8.4, there is a discussion of this particular aspect of the problem. Some important principles appear:-

a) A minimum level of discrimination in the techniques of measurement used to quantify the system and environmental changes must be established.

b) The end of the modelling curve must conform with a major discontinuity in the response of the observed system to its prescribed environmental change.

With these tools it is then possible to model field data consistently using equations of the form :-

$$Y=X^n.$$

The index *n* is then a measure of the robustness of the system. Figure 4-1 illustrates the use of this approach for a two-way spanning slab with a large hole formed off-centre and loaded uniformly. The load/deformation characteristic can be closely modelled by the equation:-

$$Y^{0.35}=X.$$

It was easy to acquire data for this example but the situation is rather different for complete buildings. Techniques for testing complete buildings in the field are becoming available [Ellis and Littler, 1988]. Much more data will be required to build up a suitable reference for new design.

The second strategy for robust design is to include devices to allow the building to avoid carrying load. This approach is similar to fusing mechanisms in electrical networks. It is already common practice in the form of expansion joints to avoid the potential damaging effects of thermal movements. As discussed in Chapter 3, the provision of weak joints is the only way to prevent the horizontal propagation of failure in systems that would otherwise have a high degree of horizontal structural integrity. What is a reasonable size of the zones left between the weak joints must be dependent upon the particular structure and its function. However, areas of not more than two bays in each direction seems to be a good starting point for the designer. The most important lesson to learned from observations of this strategy, is that the commonly held belief that only good can come from adding extra structural material to a building will not always be true.

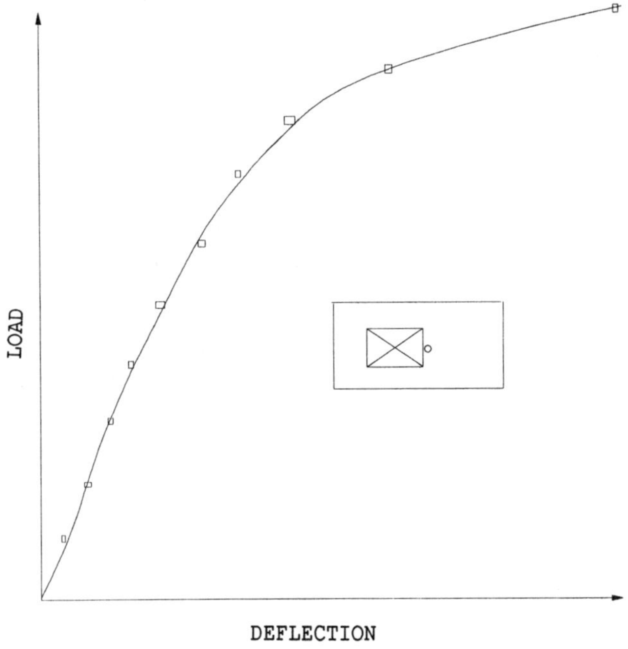

Fig. 4-1. Load deformation characteristic for two-way spanning slab

The third strategy i.e. the provision of local or global increases in resistance to enhance the overall robustness of the structure is of course applied after the normal local element design. Whatever strategies are adopted to ensure robust systems, the design of individual elements to resist the applied loads remains a necessary corner-stone of the complete design process. As indicated above care must be taken when choosing the course of overall strengthening to control robustness, since this will not necessarily change the character of the ultimate failure mechanism.

Tests at the Building Research Establishment have shown that even with the prescribed ties, it is still possible to initiate a progressive collapse in a panel structure. Nevertheless, the strategy still has most relevance to in situ concrete or precast concrete panel structures. It is less use in framed systems since in this form of construction, it is possible to lose wall and floor elements progressively and, at the same time, leave the frame essentially intact. This behaviour seems particularly possible in steel framed structures where there is normally a clear difference between the material used for the frame and that used for the infill elements.

This strategy also seems to have limited scope for application to the design of unreinforced masonry construction. It is arguable that increasing the mass in masonry buildings which in turn, increases the potential energy available to convert into the kinetic energy of a progressive failure.

Finally the fourth strategy, i.e. the installation of environmental and performance monitoring and control systems is often the most economic approach to system robustness. The siting of buildings away from potential hazards such as road traffic and the provision of protective fencing and bollards which are sacrificial will usually be much more economic than designing the building to resist large impact loads, directly. Venting of explosive loads has been discussed already and shown to be quite difficult to achieve in many instances although the use of weighted trap doors in flour and other mills has been successful in mitigating the effects of dust explosions. The best defence against gas explosions is a combination of removing as many potential sources as is possible, ensuring that there are no unventilated voids in which gas could accumulate.

Performance monitoring equipment has been installed in many important structures such as bridges, off-shore oilrigs and nuclear power stations. The reasoning behind this quite expensive exercise is that if minor changes in performance which may herald a more serious failure, can be detected electronically, then remedial action could be taken before such a hazard occurred. The classic problem here is the interpretation of the data. In the case of relatively simple structures like bridges where monitoring the location of the bridge beams relative to their bearings is often done, it is easy to specify a minimum acceptable bearing length. In the case of more complex structures the problem is far more difficult. One interesting solution has been tried in the USA for a very large roof covering an exhibition hall. In this case, a large number of strain and movement gauges were fixed to the roof structure and their output monitored during the first two years' use of the hall. The peak values obtained during this period then formed the basis for identifying significant variations in performance after that time. This still of course begs the question of what to do when the alarm bells ring but it is an ingenious development. In-service monitoring of building structures is in its infancy [Moore, 1991]. It is undoubtedly the only economic solution to the problem of degradation of structural materials which is probably the most important cause of structural failure.

5

CONTINUUM METHOD FOR REGULAR FRAMEWORKS UNDER NORMAL LOADING

The aim of this Chapter is to introduce the continuum method of large, regular frameworks. As the structures and the loads are regular, simple procedures can be developed for the stability analysis. Closed form solutions will be presented for the critical load of frameworks subjected to both concentrated forces at top level and uniformly distributed load at floor levels. The limitations and the accuracy of the formulae will also be investigated.

5.0 INTRODUCTION

The exact stability analysis of multistorey, multi-bay frameworks including all the relevant parameters is an extremely complex problem even with the help of computers. The higher the structure, the more complicated the problem is to handle. Many monographs establish the governing differential equations and outline the sometimes formidable task of solving them [Bleich, 1952; Horne and Merchant, 1965; Timoshenko and Gere, 1961]. After setting up differential equations for each element and making use of the compatibility conditions at each node, a system of differential equations is obtained [Bleich, 1952]. The equations are nonlinear, as a rule, and lead to complicated bifurcation-point problems. The same system of differential equations - and the same bifurcation-point problem - is obtained by using the energy method [Horne and Merchant, 1965]. The solution of this bifurcation-point problem results in the values of the critical loads, of which the smallest one is of practical importance. The formulation and solution of the problem in matrix form makes it possible to rely on powerful computers [Zienkiewitz, 1967; Weaver and Gere, 1990].

However, even by using computers, such analyses are fairly cumbersome to carry out, mainly because of the tedious job of preparing the input data for large structures. On the other hand, the large amount of useful and often useless output data do not help the interpretation of the results. Partly because of the above reasons and partly because of the fact that the exact solution is not really needed for practical structural engineering calculations, many simple - and not so simple - approximate methods have been developed.

To place the continuum method among the approximate methods, a background for the possible approximations shall be presented.

The first step to create an approximate method is to neglect the tertiary effect of the change in the value of the compressive forces. When the vertical load on the framework increases, the compressive forces in the members also increase but to a different extent. Due to this phenomenon, the bending stiffness of the members in compression decreases but also to a different extent, hence changing the state of stress of the framework. This change is represented in the governing equations by the non-linear terms. According to Chwalla's research [1938], this phenomenon is not important and can be neglected in most practical cases. This follows from the fact that the non-linear terms only play a significant role when the value of the external normal load is near the elastic critical load.

Having dropped the non-linear terms, the approximate procedures fall into three broad categories. Firstly, using the concept of effective length, individual columns are examined instead of the structure as a whole. The stability analysis is carried out by assessing the columns, one by one, taking into consideration the effect of the joining members as elastic restraint. The basic problem, a single column with different end-conditions, has been solved using different assumptions and is well documented [Dulácska and Kollár, 1960; Wang, 1973; Wood, 1974 and 1975]. Because of the simplicity of this approach, National Codes of Practice are often based on the effective length concept and it is widely used in design offices. The disadvantage of the method is that it cannot allow for the fact that the lateral instability of a storey is controlled by the total vertical load of the storey, rather than the load in a single column, since the stronger members support the weaker ones and the entire storey must fail as a unit [Dulácska and Kollár, 1960; Armer and Grimer, 1972; Council of Tall Buildings, 1978].

Secondly, many authors follow the concept of the exact method by proceeding step by step, carrying out successive approximations and using different simplifying assumptions in order to obtain a simpler system of differential or algebraic equations. These approximate procedures, however, often use too many simplifying assumptions and their accuracy and the range of application are limited [Bleich, 1952; Sattler, 1953; Glück et al, 1979]. In spite of this, many of the step-by-step methods are still too complicated and require electronic computation. As a rule, methods in this category do not give information about the stability of the individual elements or about the global behaviour of the structure. Hence there is an increasing demand for simpler and more economical approximate procedures by means of which a designer can make a rapid check and which enable the structural engineer to get a clear picture of the behaviour of the structure.

Thirdly, the continuum method represents an entirely different approach to the problem. The basic principle of the method is that the structure consisting of columns connected by beams is replaced by a single, *substitute* column. The bending stiffness of the *substitute* column is the sum of the bending stiffnesses of the columns of the original structure. The effect of the beams of the original structure is replaced by distributed moments acting along the substitute column. In this way the problem of the statically indeterminate structure to the n-th degree is simplified to the analysis of a statically determinate structure. The continuum method was first used by Csonka [1950] for the stress analysis of sway frames subjected to wind load. Beck [1956] applied the method to the stress analysis of Vierendeel girders. Both applications neglected the effect of the normal forces (in the columns of the framework and in the top and bottom chords of the Vierendeel girder). Both methods were later developed by the same authors to allow for the axial deformation of the members in question [Csonka, 1961; Beck, 1959]. Sigalov [1957] developed the same technique, probably not knowing of the others, for the stress analysis of one-bay asymmetric and two-bay symmetric frameworks subjected to horizontal load. The effect of the axial deformation of the columns was taken into consideration in his method.

The application of the continuum method to the linear elastic *stress* analysis of frameworks subjected to horizontal (wind) load shows that the lengthy process of calculation can be considerably reduced and closed form solution of reasonable accuracy can be derived for the maximum stresses and deformations. The continuum method has proved to be applicable to dynamic problems [Kollár, 1967; Goldberg, 1973], to coupled shear wall structures [Rosman, 1972], to sandwich columns [Hegedüs and Kollár, 1984a and 1984b], to space frames [Kollár and Hegedüs, 1985] and to buildings under abnormal loading [Goschy, 1990], as well. The application of the continuum method to different structural engineering problems is shown in Kollár's monograph [1991].

5.1 ASSUMPTIONS

When the continuum method was first applied to the stability analysis of one-bay, multistorey, regular frameworks subjected to concentrated forces on top level [Asztalos, 1972], it was suggested that the method could probably be further developed to deal with more general cases. The generalization of this "original" procedure is presented here.
During the analysis the following assumptions will be used.

a) The material of the structures is homogeneous, isotropic and obeys Hooke's law.
b) The loads are applied statically and maintain their direction during buckling (conservative forces).
c) The structures are sway structures and the critical load defines the bifurcation point.
d) The structures develop small deformations and the tertiary effect of the normal stresses in the beams is negligible. The reduction in the bending stiffness of the beams caused by the normal stresses in the beams is also negligible.
e) All storeys have the same height and the structures are at least four storeys high.
f) The structures have a rectangular network with elements attached by rigid joints to each other.
g) The stiffness (EI/l) of the beams is the same.
h) The structures have no imperfections.

5.2 SUBSTITUTE COLUMN

The continuum method of frameworks is based on the analysis of a single substitute column. The substitute column is obtained from the original framework (Fig. 5-1/a) in several steps. First, the beams are cut through at the contraflexure points (Fig. 5-1/b). Second, the shear forces at the contraflexure points are replaced with forces and moments, but along the columns (Fig. 5-1/c). These concentrated forces and moments at floor levels represent the fixing effect of the beams. These forces and moments are then uniformly distributed along the columns (Fig. 5-1/d). Finally, the columns are put together and the substitute column is created (Fig. 5-1/e). In Fig. 5-1 the columns are marked with 1, *i* and *n*, representing the 1-st, *i*-th and *n*-th column. In a similar way, the concentrated forces on the columns are marked by F_1, F_i and F_n.

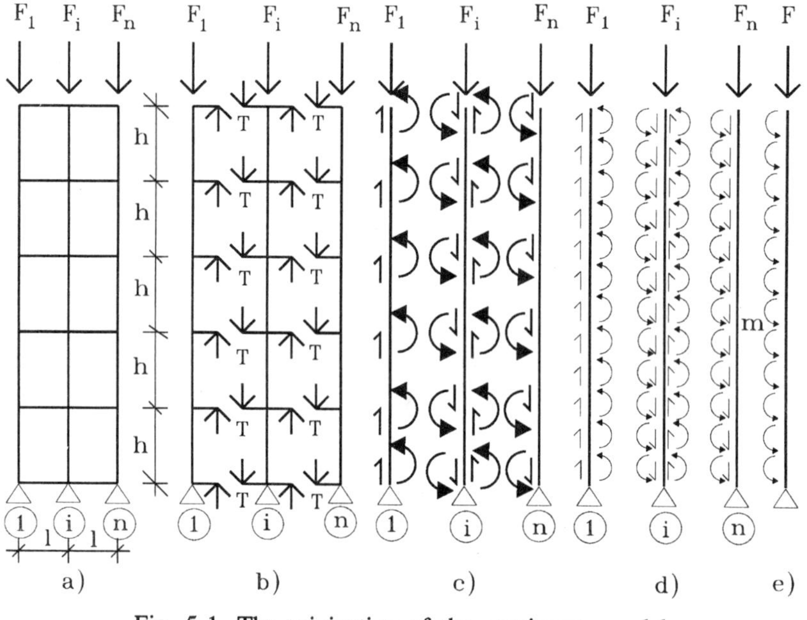

Fig. 5-1. The origination of the continuum model

The bending stiffness of the substitute column is the sum of the bending stiffnesses of the columns of the framework. The load on the substitute column equals the total load on the original frameworks. The distributed moments along the substitute column, defined precisely later on, are obtained from considerations based on the deformation of the framework. Frameworks with fixed supports lead to a cantilever as the substitute column (Fig. 5-1).

The origination of the substitute column, by putting the columns of the framework together into a single column, is justified if the columns of the framework develop the same deformation during buckling. This condition is satisfied in the case of proportional frameworks. The ratio of the stiffnesses of the beams and columns of proportional frameworks assumes certain constant values [Csonka, 1956]. A proportional framework with the characteristic ratios is shown in Fig. 5-2, where $S=EI/l$ is member stiffness and *a, b, c, d* and *e* are the constants representing the ratios.

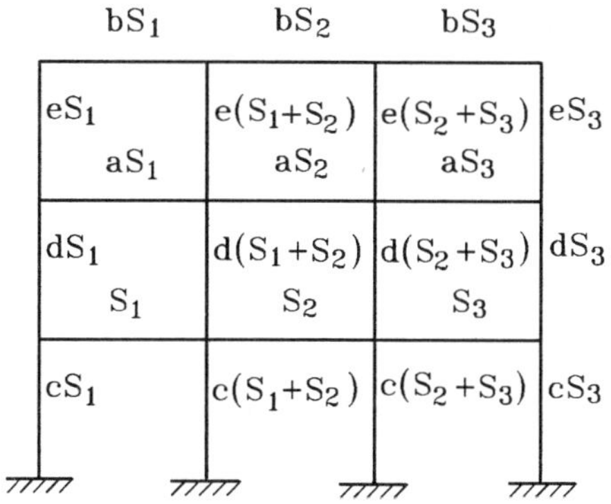

Fig. 5-2. Characteristic ratios for proportional frameworks

Figs 5-3/a and 5-3/b show practical examples for proportional frameworks with columns of constant cross–section along the height. "Proportional" in Csonka's paper [1956] referred to frameworks whose columns do not develop axial deformations.

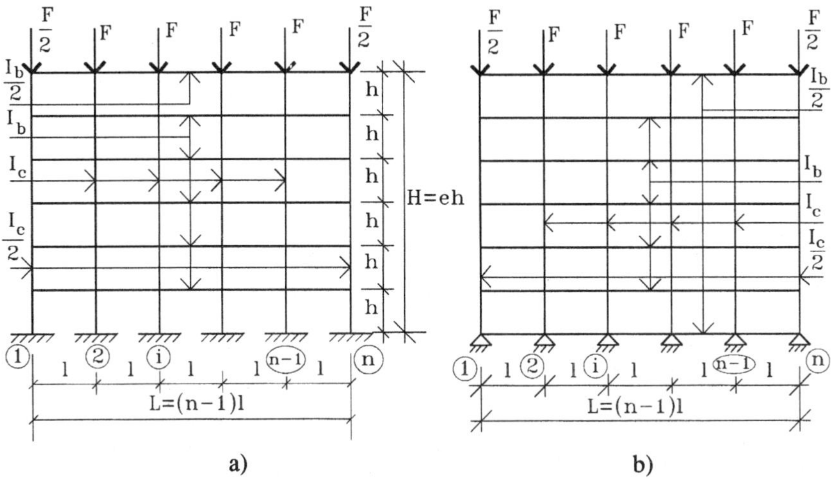

Fig. 5-3. Examples of proportional frameworks a) on fixed supports b) on pinned supports

The notion "proportionality" was generalized by Szerémi [1978] to frameworks developing column shortening and lengthening. According to his research, one-bay and two-bay symmetric frameworks automatically fulfill the conditions for proportionality but in other cases, as a rule, the conditions are hardly met.

It should be noted that when the substitute column of a framework is created, the axial forces acting along the columns (Fig. 5-1/c), resulting from the shear forces at the contraflexure points, disappear. When the columns of the framework are put together, which is equivalent to a vertical equilibrium equation, they balance each other and vanish. It follows that something is lost here: Due to the axial deformations in the columns developed by the axial forces, the actual framework develops some global sway. This phenomenon is not present in our substitute column model.

Another approximation is built into the substitute column, or more precisely, into the external load on the model. The beams develop their fixing effect on the columns at floor levels. The columns of the actual framework develop local sway (double curvature bending) between the beams. As the concentrated fixing effect of the beams was uniformly distributed when the substitute column was created, this phenomenon was also lost. These approximations will have to be dealt with later on.

5.3 FRAMEWORKS ON FIXED SUPPORTS

In addition to conditions a)...h) in Section 5.1, it is assumed that the structures investigated in this section are proportional. All the columns are of equal length, with the same stiffness, apart from the side columns whose stiffness is half of the others. All the beams are of equal length, with the same stiffness, apart from those at top level, whose stiffness is half of the others. A framework with such characteristics is presented in Fig. 5-3/a.

Although total proportionality is assumed for the correct derivation of the governing differential equations, it is clear that not all the strict conditions of proportionality are of equal importance. According to the results of a comprehensive accuracy analysis in Section 5.6, the continuum method proves to be reliable for frameworks in most practical cases. These cases cover frameworks of rectangular network with columns of the same length and cross-section and beams of the same length and cross-section. Such frameworks are defined as regular frameworks.

As is shown in Chapter 6, even structures which do not comply with the conditions of regularity can be investigated by the continuum method.

5.3.1 Frameworks Subjected to Concentrated Forces at Top Floor Level

The proportional frameworks in this section are subjected to proportional concentrated forces at top floor level. Proportionality with these forces means that they are proportional to the stiffnesses of the columns, i.e., all the forces are of the same magnitude, except for the two side forces whose magnitude is half of the internal ones (Fig. 5-3/a):

$$F_1 = F_n = F/2 \quad \text{and} \quad F_2 = F_3 = \ldots = F_{n-1} = F \, .$$

First, the governing differential equation of the substitute column will be established and solved and formulae of closed form will be produced for the stability analysis, neglecting the effect of the double curvature bending between the beams and the change in length of the columns during buckling. Second, these two phenomena will be taken into account and simple, closed form formulae will be presented for the calculation of the critical load of large frameworks on fixed supports, subjected to concentrated forces at top floor level. Frameworks with columns of variable cross-section will also be investigated.

5.3.1.1 The Continuum Model

The first step in establishing the substitute column is that the beams are cut through at the contraflexure points and the lack of continuity is replaced by shear forces T (Fig. 5-4).

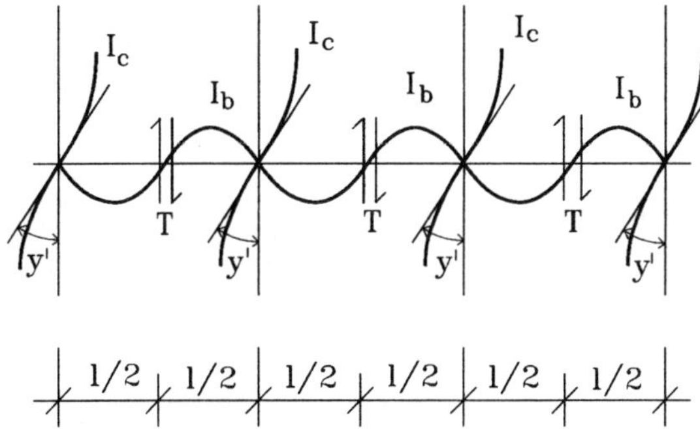

Fig. 5-4. Shear forces at the contraflexure points

The magnitude of these shear forces is obtained from the condition that the vertical translation of the contraflexure point is zero. For beams at intermediate storey levels, this condition is met if the equation

$$-y'\frac{l}{2} + \frac{T(l/2)^3}{3E_bI_b} = 0 \qquad (5.1)$$

holds, where

T is the shear force,

l is the distance between the axis of the columns,

E_b is the modulus of elasticity of the beams,

I_b is the second moment of area of the cross-sections of the beams and

$y'=dy/dz$ is the first differential quotient (Fig. 5-5/a).

The shear force is obtained from Eqn (5.1):

$$T = \frac{12E_b I_b}{l^2} y'. \tag{5.2}$$

As the next step, the shear forces at the contraflexure points are transferred to the axis of the columns. Apart from these forces, which now act as normal forces along the columns, the concentrated moments

$$M = T\frac{l}{2} = \frac{6E_b I_b}{l} y' \tag{5.3}$$

should also act on the columns. These moments and forces are then distributed uniformly along the storey height. This step is equivalent to transforming the finite problem into a continuous one. In this way, the fixing effect of the beams on the columns is expressed by the uniformly distributed moments

$$m = \frac{M}{h} = k_b y', \tag{5.4}$$

where h is the storey height. The bending stiffness of the beams, distributed along the storey height, was also introduced in Eqn (5.4):

$$k_b = \frac{6E_b I_b}{l\,h}. \tag{5.5}$$

The deflection curve of the columns is identical - this is automatically ensured by proportionality - and so the governing differential equations of the columns are also the same [Zalka, 1976]. It follows that these differential equations can be incorporated into a single equation, i.e., the columns can be put together into a single substitute column (Fig. 5-1, Fig. 5-5/a). This is the step when the normal forces along the columns and originated from the shear forces in the contraflexure points of the beams vanish. The equilibrium of an elementary section (Fig. 5-5/b) of the substitute column is represented by the equation

$$-(M + \mathrm{d}M) + M - m\mathrm{d}z + F\mathrm{d}y = 0.$$

After differentiation, making use of Eqs (5.4) and (5.5) and substituting $-E_c I_c y''$ for M, this equilibrium equation results in the governing differential equation of the substitute column, representing the framework, as

$$y'''' + \frac{F - K_b}{E_c I_c} y'' = 0\ . \tag{5.6}$$

In Eqn (5.6)

$$F = \sum_{1}^{n} F_i \tag{5.7}$$

which is the sum of the applied vertical load,

$$I_c = \sum_{1}^{n} I_{c,i} \tag{5.8}$$

which is the sum of the second moment of area of the columns and

$$K_b = 2 \sum_{1}^{n-1} k_{b,i} \tag{5.9}$$

which is the sum of the distributed bending stiffnesses of the beams, where n is the number of columns and $k_{b,i}$ is the bending stiffness of the i-th beam, defined by formula (5.5).

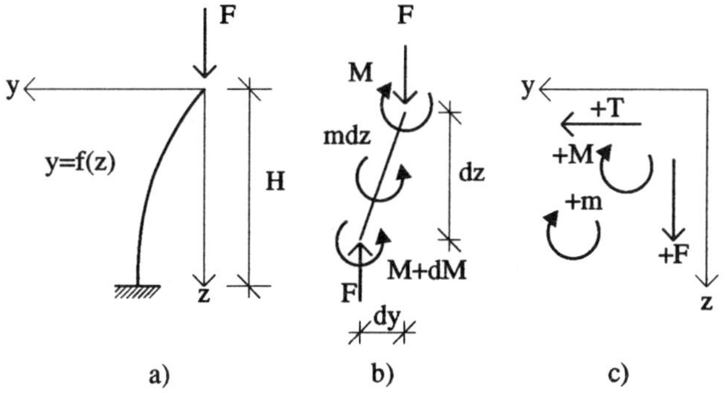

Fig. 5-5. Substitute column
a) bending shape, b) elementary section, c) sign convention

The sign convention used for the derivation of Eqn (5.6) is given in Fig. (5-5/c). The origin of the coordinate system is placed and fixed at the top of the substitute column. The boundary conditions for the substitute column in this coordinate system, when the coordinate system moves together with the top of the column during buckling (Fig. 5-5/a), are as follows.

The horizontal translation of the top of the column is zero:

$$y(0) = 0. \tag{5.10a}$$

The tangent to the column at the bottom is parallel with axis z:

$$y'(H) = 0. \tag{5.10b}$$

No moments develop at the top of the column:

$$y''(0) = 0. \tag{5.10c}$$

The sum of the shear forces at the bottom of the column is zero:

$$y'''(H) = 0. \tag{5.10d}$$

In the above boundary conditions, H represents the location of the bottom of the column, i. e., the height of the substitute column.

By making use of the above boundary conditions, the solution of the simple eigenvalue problem (5.6) is assumed in the form (Fig. 5-5/a)

$$y = f(z) = \sin\frac{\pi z}{2H}$$

and the closed formula

$$F_{cr} = F_1 + K_b \tag{5.11}$$

is found for the critical load.

In Eqn (5.11)

$$F_{cr} = \sum_1^n F_{cr,i} \tag{5.12}$$

which is the total critical concentrated load of the framework and

$$F_1 = \frac{\pi^2 E_c I_c}{4H^2} \tag{5.13}$$

which is the sum of the critical loads of the individual columns.

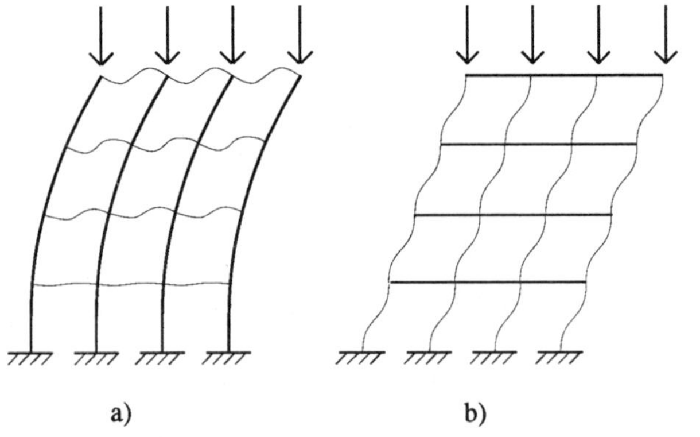

Fig. 5-6. Characteristic a) bending b) shear deformation

The two terms on the right-hand side of Eqn (5.11) are of different nature. The first term (F_l) basically depends on the height and the bending stiffness of the columns of the framework and it is associated with the bending deformation of the individual columns of height H. This part of the critical load is called the *local bending* critical load.

The second term in Eqn (5.11) is K_b, the distributed bending stiffness of the beams, defined by Eqn (5.9). It is also considered to be the shear stiffness of the framework, considered as a whole, and corresponds to the shear deformation of the structure. This part of the critical load is called the *global shear* critical load. The two types of deformation are illustrated in Fig. 5-6, bearing in mind that the substitute column represents a framework.

Formula (5.11) can be produced in a different way. Southwell's theory [Chapter 8] makes it possible to produce the critical load of a system with two different stiffnesses in a simple way. Assuming first that only one of the stiffnesses exists and then that only the other one, two part critical loads are obtained. The resulting critical load is then obtained by producing the sum of the part critical loads. The stiffness of the beams of the framework shown in Fig. 5-6/a tends to zero. The framework develops bending type deformation and the critical load practically equals the sum of the critical loads of the columns. This part critical load is F_l. If the stiffness of the columns tends to zero, the framework develops shear type deformation (Fig. 5-6/b) and the critical load is determined by the shear stiffness of the structure, i.e., the stiffness of the beams. This time the part critical load is K_b. The sum of these part critical loads is the resulting critical load of the system, according to Southwell's additive theory. This time this critical load is identical to the one obtained earlier, derived from the equilibrium of the elementary section of the structure (Eqn 5.11).

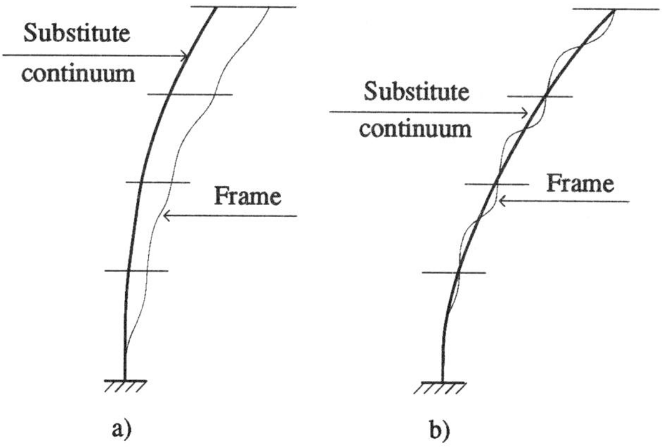

Fig. 5-7. Substitute column a) without, b) with local sway

Formula (5.11) is the exact solution to eigenvalue problem (5.6), representing the substitute column. However, it is not the exact solution, as far as the framework is concerned. The distribution of the concentrated fixing effect of the beams on the columns at floor levels results in a subtle difference between the deformation of the framework and that of the substitute continuum. During buckling, the columns of the actual framework develop additional deformation between the beams as they are only fixed by the beams at floor levels. This deformation is the double curvature bending of the columns between the beams. This double curvature bending is not present in the continuum model since the concentrated fixing effect was distributed along the height of the substitute column. Figure 5-7/a represents the substitute column without the local sway ($F_{cr}=F_l$) and Fig. 5-7/b shows the situation when the local sway is built into the model but its effect is continuously distributed ($F_{cr}=F_l+K_b$). It should be noted that Fig. 5-7/b shows exaggerated double curvature bending between the beams.

The continuum model offers a good approximation for the rotations at floor levels but it underestimates the lateral translations because only the global part of the total translations are taken into account. The lateral translations in question are associated with the stiffness of the beams, i.e., with the shear stiffness of the model. Smaller translations, in other words, greater shear stiffness was built into the continuum model. It follows that its application results in a critical load greater than the exact one.

To obtain better approximation, the affect of the additional sway of the columns should be taken into account. This will be presented now.

Every column develops additional sway between the beams (Fig. 5-8/a) but the columns on the same storey level help each other and consequently they can only fail together [Dulácska and Kollár, 1960; Armer, 1983].

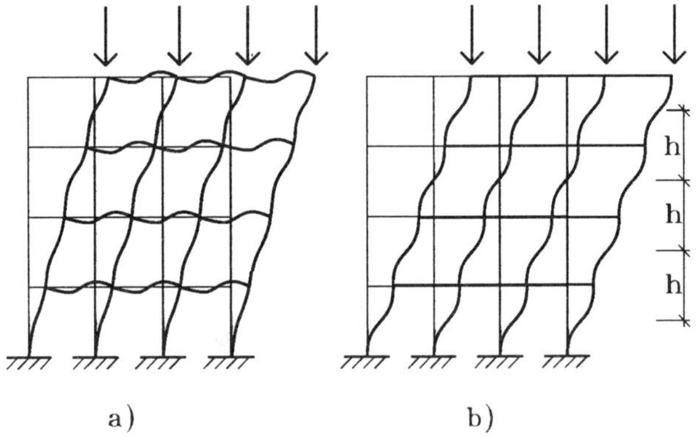

Fig. 5-8. Local sway with a) flexible, b) stiff beams

Greater sway goes with more flexible columns and stiffer beams. The situation is the most dangerous when the framework has beams of infinitely great stiffness (Fig. 5-8/b). Every storey behaves in the same way and the phenomenon is equivalent to the buckling of a system of columns of one storey high, rigidly fixed at both ends. The critical load of such a system is called the *local shear* critical load and assumes the form

$$K_c = \sum_1^n k_{c,i} = \sum_1^n \frac{\pi^2 E_c I_{c,i}}{h^2} = \frac{\pi^2 E_c I_c}{h^2}, \tag{5.14}$$

where $k_{c,i}$ is the critical load for the *i*-th column and h is the storey height. Now, when only the additional sway is considered, parameter h is also the effective length of the columns in question. The effective length of the columns of a one storey system and the critical load of the system are always in close connection when the additional sway is investigated so that we shall rely on the effective length for the determination of the critical load of the additional sway for frameworks on different supports as well.

The second term (K_b) in Eqn (5.11) and K_c given by Eqn (5.14) both represent critical loads and they are both associated with shear deformation. Suppose first that the columns of the framework are restricted from bending between the beams. The resistance of the framework through shear deformation is characterized by K_b (Eqn 5.11). Second, suppose that the beams of the framework are restricted from bending. The framework develops shear deformation again and the corresponding critical load is defined by K_c (Eqn 5.14). On the basis of the Föppl-Papkovics theorem [Chapter 8], these part critical loads can be combined. The reciprocal addition

$$\frac{1}{K} = \frac{1}{K_b} + \frac{1}{K_c} \tag{5.15}$$

results in

$$K = K_b \frac{K_c}{K_c + K_b}, \tag{5.16}$$

where K is the shear stiffness of the framework, including the effects of the additional local sway of the columns. K is also called the *shear* critical load. As both part critical loads represent the same type of deformation, Eqn (5.16) is considered exact, as far as shear deformation is concerned. According to formula (5.16), the additional local sway can be interpreted as a phenomenon which reduces the shear stiffness of the framework, defined earlier as K_b in (5.9). The reduction factor is

$$\frac{K_c}{K_c + K_b}.$$

It is interesting to note that a more detailed analysis based on the equilibrium of the nodes of the framework results in the same formula for the shear stiffness of the framework [Zalka, 1977b; Kollár, L. P. 1986].

The critical load of the continuum model, including the effects of local sway, can now be presented. All that needs to be done is the replacement of K_b in Eqn (5.9) with K. By so doing, the formula

$$F_{cr} = F_1 + K, \tag{5.17}$$

is obtained, where F_1 represents the critical load of the columns of the framework, acting as individual cantilevers in bending (Eqn 5.13), and K is the shear stiffness of the framework, i.e., the shear critical load.

The effect of the additional sway of the columns is now built into the continuum model but formula (5.17) still does not take into consideration the axial deformation of the columns during buckling. Before the effect of the axial deformation is investigated in the next Section, it is only mentioned that with building the effect of the additional sway of the columns between the beams into the continuum model, we have not eliminated the whole approximation resulting from the distribution of the concentrated fixing effect of the beams. The fact that the fixing effect was distributed *downwards* is in itself an approximation which increases the value of the critical load. This slight approximation, however, will remain in the continuum model.

5.3.1.2 Axial Deformation of the Columns

When the columns were "added up" into a substitute column (Fig. 5-1), the vertical forces (small arrows in Fig. 5-9) resulting from the shear forces at the contraflexure points of the beams were also added up in a fictitious vertical equilibrium equation. They balanced each other and vanished [Zalka, 1976]. Because of them, however, the columns of the actual framework may undergo different axial deformations, i.e., shortening on one side of the framework and lengthening on the other one. Figure 5-9/a concentrates on the axial deformation of the columns and does not show the double curvature bending of the beams and columns during buckling. Figure 5-9/b includes the double curvature bending of the beams but the bending of the columns between the columns are still not shown.

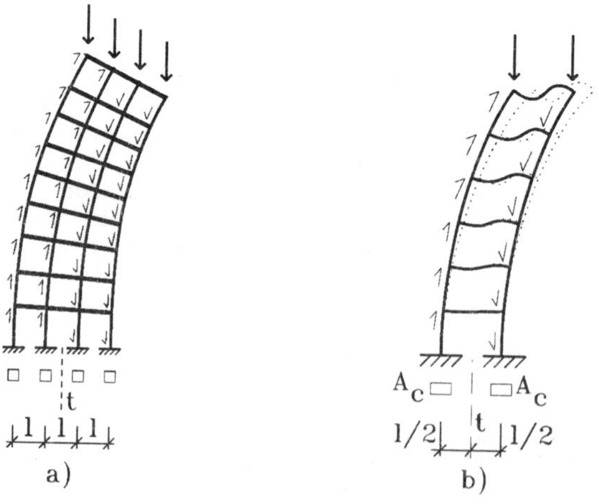

Fig. 5-9. Additional global buckling of a) multi-bay and b) single-bay frameworks

Due to the change in length of the columns, the beams undergo bending and the framework softens up. Another additional sway occurs, but this time the framework as a whole develops additional sway. This phenomenon can be interpreted as the additional *global* buckling of the framework - as opposed to the *local* sway discussed above. If the framework is regarded as a bar whose longitudinal fibres are the columns of the framework, then the deformation now can be characterized as bending deformation. In this case the *global* bending stiffness of the framework is represented by the second moment of area of the cross-sections of the columns, with respect to the centroidal axis of the cross-sections of the columns (this axis is marked with t in Fig. (5-9/a).

The phenomenon is clearly seen with one-bay, proportional frameworks [Asztalos, 1972]. The softening up of the structure results from the difference in the translations of the beam-ends. This difference is directly caused by the forces acting along the columns. These forces have come from the shear forces at the contraflexure points of the beams and make the columns develop shortening on one side and lengthening on the other side. This axial deformation then makes the structure develop global sway (Fig. 5-9/b). After solving a system of differential equations, Asztalos [1972] found a solution of closed form for the critical load of one-bay, multistorey, proportional frameworks subjected to concentrated forces at top floor level. After considerable rearrangement and with notation introduced in this chapter, his formula can be rewritten as

$$F_{cr} = \frac{F_1 + K + \dfrac{F_1 K}{F_g^*}}{1 + \dfrac{K}{F_g^*}}, \qquad (5.18)$$

where

$$F_g^* = \frac{\pi^2 E_c 2A_c (l/2)^2}{4H^2} \qquad (5.19)$$

is the part critical load which belongs to the additional global sway. In formula (5.19) A_c and E_c are the cross-sectional area and the modulus of elasticity of the columns, respectively. The term

$$2A_c\left(\frac{l}{2}\right)^2$$

in the numerator of formula (5.19) represents the second moment of area of the cross-sections of the columns to the centroidal axis of the cross-sections (marked with t in Fig. 5-9/b). Formula (5.19) only holds for one-bay frameworks but it can be generalized. The second moment of area of the cross-sections of the columns with multi-bay frameworks (Fig. 5-10) is expressed by

$$I_g = \sum_1^n A_{c,i} t_i^2 \tag{5.20}$$

where $A_{c,i}$ and t_i are the cross-section of the *i*-th column and its distance from the centroidal axis, respectively. With the *global* inertia I_g, the *global bending* critical load of multi-bay frameworks, representing the additional global sway, is

$$F_g = \frac{\pi^2 E_c I_g}{4H^2} \tag{5.21}$$

and the critical load of the framework assumes the form

$$F_{cr} = \frac{F_l + K + \dfrac{F_l K}{F_g}}{1 + \dfrac{K}{F_g}} = F_l + \left(\frac{1}{F_g} + \frac{1}{K}\right)^{-1}. \tag{5.22}$$

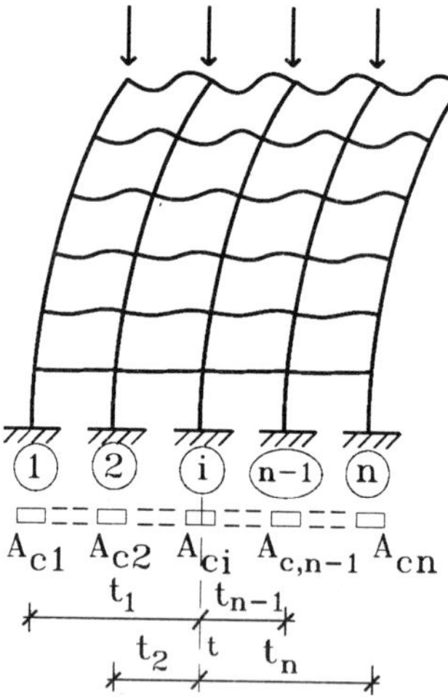

Fig. 5-10. Global bending sway of multi-bay framework

According to the results of numerical analyses, the effect of the local sway (additional sway of the columns between the beams) is considerable, as a rule, and is always to be taken into account. On the other hand, the effect of the global sway (additional sway of the framework as a whole) is only significant with tall, slim frameworks.

Formulae (5.20) and (5.21) show that the effect of the axial deformation of the columns increases in direct ratio to the decrease of the area of the columns and in quadratic ratio to the increase of the height of the framework and the decrease of the distance between the columns.

It is interesting to note that formula (5.22) can also be produced by the repeated application of Southwell's additive formula and the Föppl-Papkovics reciprocal additive formula.

When analysing multi-layer sandwich cantilevers with thick faces, Kollár derived the same global stiffness EI_g as appears in formula (5.21) [Hegedüs and Kollár, 1984b]. He also demonstrated that for one-bay and two-bay symmetrically loaded frameworks, the formula is exact. In the case of frameworks with more than two bays, the formula is approximate but with nine-bay frameworks the error becomes negligible. With medium-rise frameworks with four or more bays, the effect of the additional global sway is not significant anyway.

The results of a comprehensive numerical analysis show that formula (5.22) sometimes overestimates the critical load. The reason of this overestimation is that the stiffness of the columns is taken into account twice: Once in the local critical load F_l and then again in the local shear stiffness K_c. A sophisticated mathematical investigation shows that the two phenomena cannot develop in full depth at the same time [Hegedüs and Kollár, 1987]. By introducing the reduction factor

$$r = \frac{K_c}{K_c + K_b} \tag{5.23}$$

into formula (5.22), this discrepancy can be corrected. By doing so, the formula for the critical load finally emerges as

$$F_{cr} = \frac{rF_l + K + \dfrac{rF_l K}{F_g}}{1 + \dfrac{K}{F_g}} = rF_l + \left(\frac{1}{F_g} + \frac{1}{K} \right)^{-1} . \tag{5.24}$$

Factor r, ranging between zero and one, is also useful to characterize the behaviour of the framework, as being dominated by the stiffness of the columns or by the stiffness of the beams. The factor tends to one in the first case and tends to zero in the second case.

5.3.1.3 Frameworks with Variable Cross-sectioned Columns

In order to decrease the weight of frameworks, columns with variable cross-section are sometimes used. Consider again the framework introduced in Section 5.3.1, but this time let the second moment of cross-section of the columns vary:

$$I_{c,i} = I^u_{c,i}\left(\frac{z + d}{d}\right)^2 . \qquad (5.25)$$

The continuum model of the framework is developed according to the procedure presented in Section 5.3.1.1 (Fig. 5-11). The second moment of area of the cross-section of the column assumes the value I^u_c at the top and I^b_c at the bottom of the column. The distance between the top of the column and the location where the second moment of area of the cross-section would assume zero value is defined by parameter *d*.

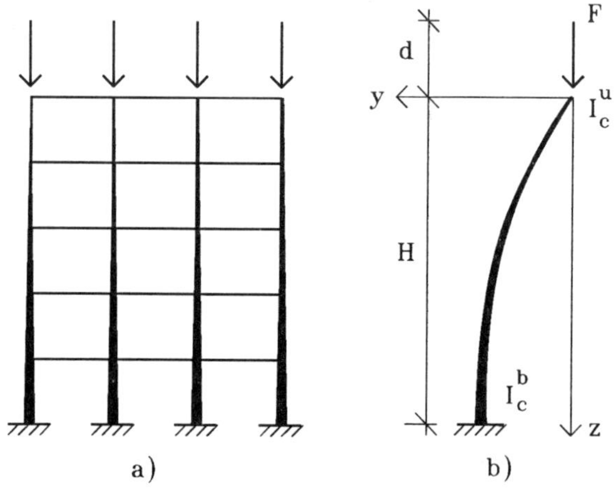

Fig. 5-11. Frameworks with columns of variable cross-section
a) theoretical model and b) substitute column

Examining the equilibrium of an elementary section of the substitute column, the four-order differential equation of variable coefficients

$$\left(z^2 y''\right)'' + \frac{(F - K_b)d^2}{E_c I^u_c} y'' = 0 \qquad (5.26)$$

is derived as the governing differential equation of the continuum model [Zalka, 1976]. The boundary conditions for this equation coincide with those established for the frameworks with columns of constant cross-section: Eqs (5.10a/b/c/d). The only new term in Eqn (5.26) is the sum of the second moment of area of the cross-section of the columns at the top of the framework:

$$I^u_c = \sum_1^n I^u_{c,i} . \qquad (5.27)$$

After reducing the order of differential equation (5.26) twice, it assumes a form identical to that of a cantilever of variable cross-section, subjected to a concentrated force - our problem can be considered as the generalization of that of the cantilever. The solution of the cantilever is readily available [Timoshenko and Gere, 1961] and its application to our problem, also using the notation introduced in this Chapter, results in

$$F_{cr} = g\bar{F}_1 + K_b \tag{5.28}$$

for the critical load of the continuum model of the framework with columns of variable cross-section. The local critical load of the columns is obtained from

$$\bar{F}_1 = \frac{E_c I_c^b}{H^2}, \tag{5.29}$$

where I_c^b is the sum of the second moment of area of the cross-section of the columns at the bottom of the framework. Global shear stiffness K_b is defined by formulae (5.9) and (5.5). Values for parameter g are presented in Table 5-1 in the function of the ratio I_c^u/I_c^b.

Table 5-1. Values for parameter g

I_c^u/I_c^b	0.0	0.1	0.2	0.3	0.4	0.5	0.6	0.7	0.8	0.9	1.0
g	0.250	1.350	1.593	1.763	1.904	2.023	2.128	2.223	2.311	2.392	$\pi^2/4$

Formulae (5.28) and (5.11) are similar to each other not only in form but also in the fact that neither takes into consideration the effect of the local shear and global bending sway.

As far as the additional sway of the columns between the beams is concerned, the upmost storey region is to be considered since it is there where the cross-section of the columns is the smallest, in other words, where their resistance against local buckling is minimum.

The cross-section of the columns turned out to be an important factor when the effect of the axial deformation of the columns were investigated in Section 5.3.1.2. The smaller the cross-section of the columns, the greater the effect of the axial deformation of the columns.

Both the local and global sway are of "secondary" importance. Bearing this in mind and the fact that the exact analysis of these effects would be very difficult, a compromise is suggested. Take both effects into consideration on the basis of formulae (5.24) and (5.28) but when the "secondary" effects are considered, assume that the cross-sections of the columns are constant, according to the smallest cross-section, i.e., the cross-section at the top of the structure. By doing so, we arrive at

$$F_{cr} = \frac{r g \bar{F}_1 + K + \dfrac{r g \bar{F}_1 K}{F_g}}{1 + \dfrac{K}{F_g}} = r g \bar{F}_1 + \left(\frac{1}{F_g} + \frac{1}{K} \right)^{-1} . \qquad (5.30)$$

For shear stiffness K and global critical load F_g in formula (5.30), expressions (5.16), (5.21) and (5.20) can be used but with the minimum cross-section, i.e., with I_c^u and A_c^u.

5.3.2 Frameworks with UDL at Floor Levels

Proportional frameworks whose load is a uniformly distributed load of intensity p along the beams (Fig. 5-12/a) will be dealt with in this section. In two steps, the UDL on the beams of the framework can be approximately replaced by a continuous load system which is uniformly distributed along the columns (Fig. 5-12/b/c).

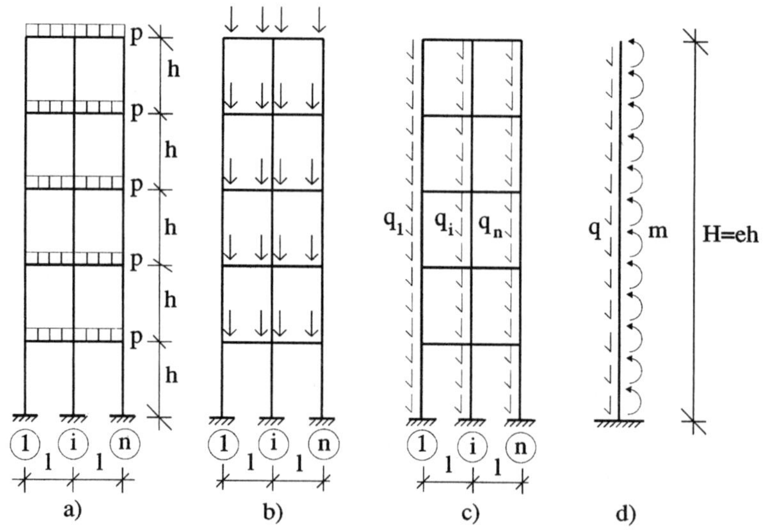

Fig. 5-12. The origination of the continuum model

First, the UDL on the beams is transferred to the columns at storey levels (Fig. 5-12/b) in the form of concentrated forces (the reactions on the beams). These concentrated forces are then distributed along storey height (Fig. 5-12/c), in fact along the height of the framework. The beams are cut through at contraflexure points (not shown in Fig. 5-12 but in Fig. 5-1) and finally the columns are added up into a single substitute cantilever (Fig. 5-12/d), as it was explained in details in Section 5.2.

The relationship

$$N = pLe = qHn \tag{5.31}$$

holds between the original load of the framework defined by intensity p and the "substitute" load characterized by intensity q. In Eqn (5.31) e is the number of storeys and N is the total vertical load of the framework measured at ground floor level.

The above establishment of the substitute load results in an unfavourable approximation. It is only possible to replace the UDL on the beams by UDL along the entire length of the columns if the concentrated forces at floor levels, resulting from the load on the beams, are distributed *downwards* along the columns (Fig. 5-12/b/c). In this way, the centre of the vertical load of the substitute cantilever is in a lower position than with the actual framework and this leads to an overestimated critical load. A formula to reduce the error is presented in Section 5.5. Exactly in the same way as with frameworks subjected to concentrated forces (Section 5.3.1.1), the fixing effect of the beams at the storey levels is distributed *downwards*. This is also an unfavourable approximation.

Another phenomenon should be mentioned here. The magnitude of the vertical load of the framework and also of the substitute column increases going downwards from the top of the structure to ground floor level. The second moment of area of the cross-section of the columns and also of the substitute column, however, is constant along the height of the structure. This again results in slight overestimation of the critical load.

5.3.2.1 The Continuum Model

The governing equation of the substitute column of frameworks subjected to UDL on the beams is the fourth order differential equation of variable coefficients

$$y'''' + \left(\frac{N(z) - K_b}{E I_c} y' \right)' = 0 , \tag{5.32}$$

where

$$N(z) = \sum_1^n N_i(z) = zq = z \sum_1^n q_i \tag{5.33}$$

is the vertical total load at height z and q is the intensity of the vertical load along the substitute column [Zalka, 1977a]. Parameters I_c and K_b are defined by formulae (5.8) and (5.9).

The origin of the coordinate system is fixed at the top of the substitute column so that the boundary conditions of the problem are as follows.

The lateral translation of the top of the column is zero:

$$y(0) = 0. \tag{5.34a}$$

The tangent to the substitute column at the bottom of the column is parallel to axis z:

$$y'(H) = 0. \tag{5.34b}$$

No bending moment develops at the top of the column:

$$y''(0) = 0. \tag{5.34c}$$

The sum of the shear forces at the bottom of the column assumes zero value:

$$y'''(H) = 0. \tag{5.34d}$$

By integrating Eqn (5.32) once and making use of boundary condition (5.34d), the order of the differential equation can be reduced by one:

$$y''' + \frac{N(z) - K_b}{E_c I_c} y' = 0. \tag{5.35}$$

Boundary conditions (5.34a/b/c) go with this differential equation.

The third order, homogeneous differential equation of variable coefficient has no solution of closed form and the widespread numerical methods fail in certain stiffness regions. To overcome the numerical problems and also to get a picture of the response of the structure to buckling, apply Southwell's additive theorem [Chapter 8] to obtain a lower bound to the lowest eigenvalue, i.e., to the critical load of the framework. This procedure not only yields a lower bound which can be efficiently used for the determination of the exact critical load but also shows the effect of the different stiffness characteristics.

According to Southwell's theorem, consider first the bending stiffness of the substitute column ($E_c I_c \neq 0$) only and disregard the shear stiffness of the substitute column which is in fact the bending stiffness of the beams (K_b=0). In this case Eqn (5.35) assumes the form

$$y''' + \frac{N(z)}{E_c I_c} y' = 0. \tag{5.36}$$

This differential equation, together with boundary conditions (5.34a/b/c) represent a cantilever subjected to gravity load. The equation can be traced back to the Bessel-type differential equation and, making use of Bessel functions, the closed form solution

$$N_1 = \frac{7.837 E_c I_c}{H^2} \tag{5.37}$$

is obtained [Timoshenko and Gere, 1961] as the part critical load of the substitute cantilever. The critical load is defined as the maximum value of the normal load at ground floor level. The deflection curve representing pure bending deformation is presented in Fig. 5-13.

Second, assume that the substitute column only resists the external load by making use of its shear stiffness, i.e., the distributed stiffness of the beams, and disregard the bending stiffness ($E_c I_c$=0). Equation (5.35) now simplifies to

$$y'\left(N(z) - K_b\right) = 0, \qquad (5.38)$$

where $N(z)=qz$. Boundary condition (5.34a) belongs to Eqn (5.38). This differential equation represents the eigenvalue problem of a column which only develops shear deformation. The part critical load of the substitute column will be obtained through both mathematical and physical considerations.

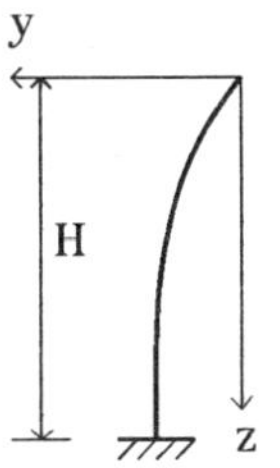

Fig. 5-13. Bending deformation

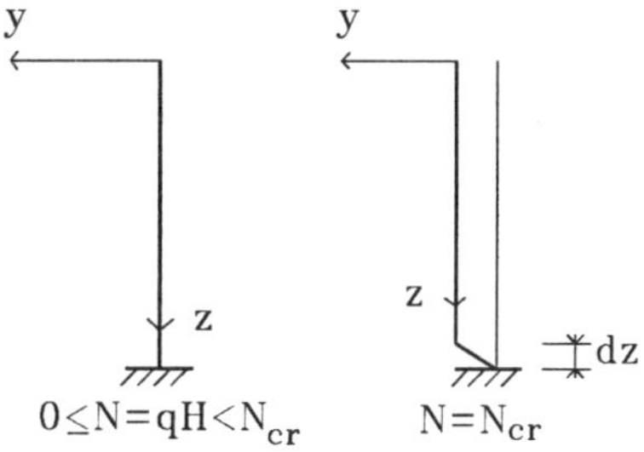

Fig. 5-14. Shear deformation

The left hand side of equation (5.38) represents a product and a product can assume zero value only if one of its terms vanishes. Equation $y'=0$ cannot hold along the *whole* column, because in this case, by virtue of condition $y(0)=0$, there would be no deflection at all. However, equation $y'=0$ can hold in the open domain $0 \le z < H$ while equation $qz-K_b=0$ can hold at one point at the open domain $0 < z \le H$. Bearing in mind that only the lowest possible eigenvalue (the critical load intensity q_{cr}) is needed, it turns out that equation $qz-K_b=0$ yields the lowest eigenvalue at $z=H$, i.e., we obtain

$$q_{cr} = \frac{K_b}{H}$$

from which the part critical load representing the shear resistance of the column emerges as

$$N_{cr} = q_{cr} H = K_b \ . \qquad (5.39)$$

Accordingly, equation $y'=0$ does hold in the open domain $0 \le z < H$ and then the second term in Eqn (5.38) holds at $z=H$. The corresponding eigenfunction is unique: The axis of the column develops a sudden deflection at $z=H$, otherwise it remains straight (Fig. 5-14).

Having determined the two part critical loads, a lower bound is now obtained for the original problem on the basis of Southwell's theorem, by producing the sum of the critical loads in Eqs (5.37) and (5.39):

$$N_{cr} = N_l + K_b \,. \tag{5.40}$$

Southwell's theorem results in the exact critical load when the component structures develop the same type of deformation. When the deformations are of different type, the theorem produces a lower bound. The magnitude of the difference between the exact solution and the approximate one depends on how much the deformations differ from each other in nature. Now, the first term in Eqn (5.40) is associated with bending deformation (Fig. 5-13) and the second term represents buckling in shear in the form of a straight line (Fig. 5-14) so that the formula results in a conservative estimate for the critical load of the substitute column. N_l is the *local bending* part critical load and K_b is defined as the *global shear* part critical load.

It is interesting to note that Eqs (5.40) and (5.11) are of the same form. However, while Eqn (5.40) is a conservative estimate, Eqn (5.11) is exact, as far as the *substitute column* is concerned. This follows from the fact that with the cantilever subjected to a concentrated force, the eigenfunction which belongs to the shear type deformation represents arbitrary shape. This shape can even assume the shape of the bending deformation which characterizes the first term in Eqn (5.11).

Knowing a conservative estimate for the critical load, it is easier to find the solution to the eigenvalue problem (5.35) by some numerical method. Equation (5.35) can be transformed to a boundary-value problem and an initial-value problem. The application of the finite difference method then leads to a reasonably stable procedure [Zalka, 1977a]. In the function of the stiffness ratio

$$\beta = \frac{K_b}{N_l} \tag{5.41}$$

the critical load parameter α can be produced, with which the critical load of the substitute column is obtained from

$$N_{cr} = \alpha N_l . \tag{5.42}$$

Values for the critical load parameter α are given in Figs 5.15 and 5.16 and in Table 5-2 in the function of the stiffness ratio β. Formulae (5.40) and (5.42) make it possible to check the accuracy of Southwell's theorem. The critical load was computed by using Southwell's approximate formula (5.40) and by the exact formula (5.42) for different stiffness ratios β and the deviation (Δ) of the approximate value from the exact one was determined (Table 5-2). The comparison shows that Southwell's formula always yields a conservative estimate and that the maximum deviation is 47%. The approximate values obtained by the application of Southwell's formula are also given in Figs 5-15 and 5-16.

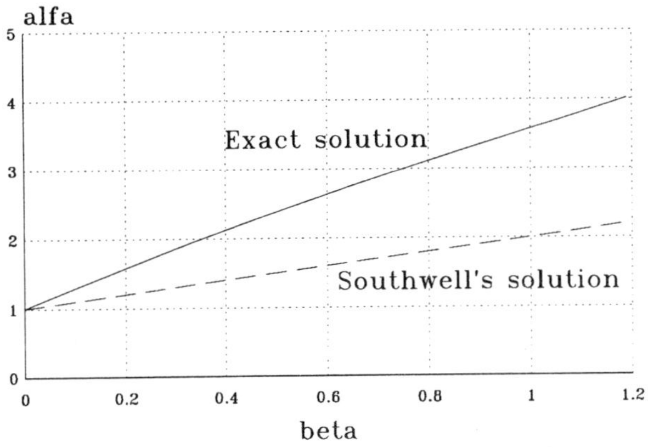

Fig. 5-15. Values for critical load parameter α

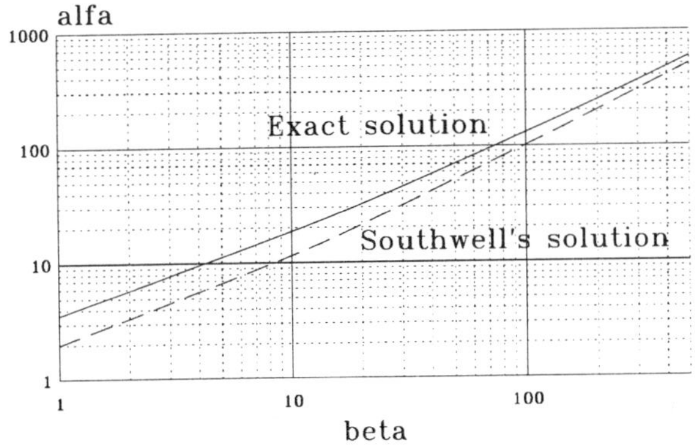

Fig. 5-16. Values for critical load parameter α

In making use of Eqn (5.41), formula (5.42) can be rearranged as

$$N_{cr} = \frac{\alpha}{\beta} K_b. \tag{5.43}$$

This formula directly shows the relationship between the critical load and the shear stiffness of the substitute column.

It is interesting to note that the in-plane buckling of cantilevers with both bending and shear stiffness, discussed here, and the pure torsional buckling of cantilevers with torsional and warping stiffnesses, to be dealt with in Chapter 7, are analogous physical problems. The analogy will be fully exploited in Chapter 7 where the formulae, tables and figures presented here will be used.

Table 5-2. Load parameter α for frameworks on fixed supports
α : continuum solution α_s : Southwell Δ : Southwell's error

β	α_s	α	Δ	β	α_s	α	Δ	β	α_s	α	Δ
0.0000	1.0000	1.0000	0	0.1	1.1	1.295	15	30	31	44.86	31
0.0005	1.0005	1.0015	0	0.2	1.2	1.580	24	40	41	57.54	29
0.001	1.001	1.0030	0	0.3	1.3	1.856	30	50	51	69.99	27
0.002	1.002	1.0060	0	0.4	1.4	2.123	34	60	61	82.27	26
0.003	1.003	1.0090	1	0.5	1.5	2.382	37	70	71	94.40	25
0.004	1.004	1.0120	1	0.6	1.6	2.633	39	80	81	106.44	24
0.005	1.005	1.0150	1	0.7	1.7	2.878	41	90	91	118.38	23
0.006	1.006	1.0180	1	0.8	1.8	3.116	42	100	101	130.25	22
0.007	1.007	1.0210	1	0.9	1.9	3.349	43	120	121	153.79	21
0.008	1.008	1.0240	2	1.0	2.0	3.576	44	140	141	177.13	20
0.009	1.009	1.0270	2	2.0	3.0	5.624	47	160	161	200.30	20
0.010	1.010	1.0300	2	3.0	4.0	7.427	46	180	181	223.33	19
0.020	1.020	1.0598	4	4.0	5.0	9.100	45	200	201	246.24	18
0.030	1.030	1.0896	5	5.0	6.0	10.697	44	300	301	359.51	16
0.040	1.040	1.1192	7	6.0	7.0	12.241	43	400	401	471.28	15
0.050	1.050	1.1487	9	7.0	8.0	13.749	42	500	501	581.79	14
0.060	1.060	1.1782	10	8.0	9.0	15.227	41	600	601	684.82	12
0.070	1.070	1.2075	11	9.0	10.0	16.683	40	700	701	765.61	8
0.080	1.080	1.2367	13	10.0	11.0	18.118	39	800	801	842.45	5
0.090	1.090	1.2659	14	20.0	21.0	31.820	34	900	901	912.66	1

As it was shown with frameworks subjected to concentrated forces in Section 5.3.1.1, the shear stiffness of the substitute column K_b defined by formulae (5.9) and (5.5) should be modified. The beams of the framework are not continuously distributed as is assumed with the continuum model, thus the columns undergo additional local sway between the beams. This was also the case with frameworks under concentrated load. The only difference is in the intensity of the external load. The intensity of the external load causing local sway was constant with frameworks subjected to concentrated forces at the top but now it increases along the height of the structure (Fig. 5-17).

Assuming beams with infinitely great stiffness as the most unfavourable case, as far as the local sway is concerned, it turns out that the first storey region has to be investigated, where the external load assumes maximum value. This is the same situation as with frameworks subjected to concentrated forces so that the results obtained in Section 5.3.1.1 can be used. Accordingly, everywhere in this section shear stiffness K_b should be replaced by K, defined by formula (5.16), and all the formulae derived not only hold but also take the effect of local sway into consideration. The formula for the determination of the critical load of the substitute column is (5.42) or, on the basis of formula (5.43)

$$N_{cr} = \frac{\alpha}{\beta} K , \tag{5.44}$$

where

$$\beta = \frac{K}{N_1} . \tag{5.45}$$

Values for parameter α are presented in Table 5-2 and in Figs 5-15 and 5-16 in the function of ratio β.

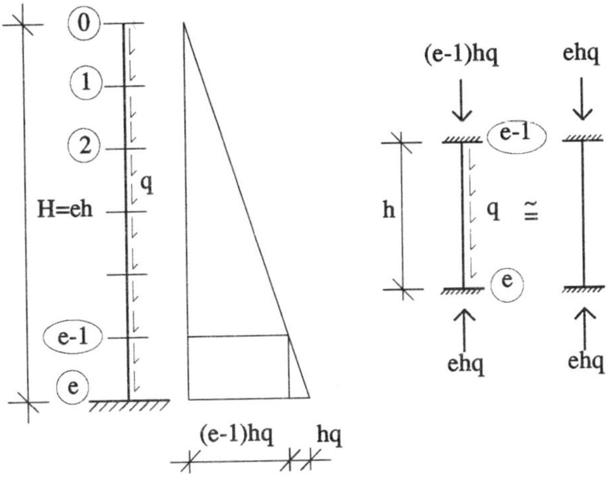

Fig. 5-17. UDL along the substitute column

The shear stiffness in the above formulae is defined by formulae (5.16), (5.14) and (5.9). Formula (5.37) yields the part critical load N_1.

Formula (5.44) is simple enough but it does not show clearly that the columns - through their local bending - and the beams - through their global shear effect - contribute separately to the resistance of the whole structure against buckling failure. If we rearrange formula (5.44), we arrive at

$$N_{cr} = (\alpha - \beta)N_1 + K . \tag{5.46}$$

This formula is still simple and also shows how the two characteristic deformations affect the critical load.

5.3.2.2 Axial Deformation of the Columns

As with frameworks subjected to concentrated forces, the unfavourable effect of the different axial deformations of the columns has to be taken into consideration. Due to the normal forces resulting from the shear forces at the contraflexure points of the beams, the columns develop lengthening and shortening of different magnitude. The distribution of these normal forces, however, is different with frameworks subjected to concentrated forces and with those subjected to UDL. A concentrated force of a given magnitude causes bigger deflection acting at the top of the bar of a given length than a distributed load of the same magnitude acting along the bar. In our case, smaller deflection of the framework subjected to UDL goes with smaller shear forces at the contraflexure points, which result in smaller axial deformation. It follows that, bearing in mind this difference and the fact that this phenomenon is only of secondary nature, with good approximation, the *structure* of the terms established for the axial deformations of the columns of frameworks subjected to concentrated forces can be borrowed. Due to the uniformly distributed nature of the load, the part critical load belonging to the additional global buckling assumes the form

$$N_g = \frac{7.837 E_c I_g}{H^2}, \qquad (5.47)$$

where I_g is the global second moment of area of the cross-sections of the columns. Its value does not depend on the nature of the load of the framework since it represents the bending resistance of the framework as a whole against buckling, through the shortening and lengthening of the columns as longitudinal fibres. It follows that we can take formula (5.20) for I_g, established in Section 5.3.1.2 with frameworks subjected to concentrated forces. With the approximate formula (5.47), all that needs to be done is to expand formula (5.46) to incorporate the effect of the axial deformation. This can easily be done on the basis of equation (5.22). By so doing, we arrive at the approximate critical load of frameworks subjected to UDL on the beams:

$$N_{cr} = \frac{(\alpha-\beta)N_l + K + \dfrac{N_l K}{N_g}}{1 + \dfrac{K}{N_g}}. \qquad (5.48)$$

Values for parameters α and β needed for the computation are to be found in Table 5-2 and in Figs 5-15 and 5-16. The part critical loads N_l, N_g and K are defined by formulae (5.37), (5.47) and (5.16), respectively.

The repeated application of Southwell's additive theorem and the Föppl-Papkovics reciprocal additive theorem [Section 8.1] now results in the lower bound

$$N_{cr} = \frac{N_1 + K + \dfrac{N_1 K}{N_g}}{1 + \dfrac{K}{N_g}} = N_1 + \left(\frac{1}{N_g} + \frac{1}{K}\right)^{-1}. \tag{5.49}$$

to the critical load. The difference between formulae (5.48) and (5.49) is in the numerator of the formulae. The term (N_1+K) in the numerator of formula (5.49) is a conservative approximation obtained by Southwell's theorem. On the other hand, the term

$$(\alpha\text{-}\beta)N_1 + K$$

in the numerator of formula (5.48) is the exact solution of the eigenvalue problem of the substitute continuum. The difference between the two terms may go up to 47%.

Although the term

$$(\alpha\text{-}\beta)N_1 + K$$

in formula (5.48) is exact and it represents the dominant part of the critical load, as a rule, formula (5.48) is not an exact formula and depending on the stiffness characteristics of the framework, it may result in a critical load which is greater than the exact one. This follows from the fact that unfavourable phenomena were neglected when the substitute column was created - see also remarks made in Section 5.3.2. Results of a comprehensive accuracy analysis is presented in Section 5.6.

Formula (5.48) based on the application of the continuum model takes several phenomena into consideration. As was previously pointed out in Section 5.3.1.2 where frameworks under concentrated forces at top floor level were discussed, all these phenomena could not develop at the same time in certain stiffness regions. To allow for this fact, modifier r defined by formula (5.23) can again be introduced and the formula for the critical load finally assumes the form

$$N_{cr} = \frac{(\alpha\text{-}\beta)rN_1 + K + \dfrac{rN_1 K}{N_g}}{1 + \dfrac{rK}{N_g}}. \tag{5.50}$$

Because of the non-uniform distribution of the UDL on the beams along the columns, there is an unfavourable approximation behind formula (5.50). As is seen in Section 5.5, the effect of this approximation can be reduced and the accuracy of formula (5.50) can be further improved.

5.4 FRAMEWORKS ON PINNED SUPPORTS

Apart from assumptions a)...h) in Section 5.1, the frameworks on pinned supports investigated here also have to fulfill the following criteria:

1) All the columns are of the same length with the same stiffness, except for the side columns whose stiffness is half of the inside ones,

2) The length of all the beams is the same,

3) The stiffness of the beams is the same, except for the beams at the top and ground floor levels, whose stiffness is half of the others.

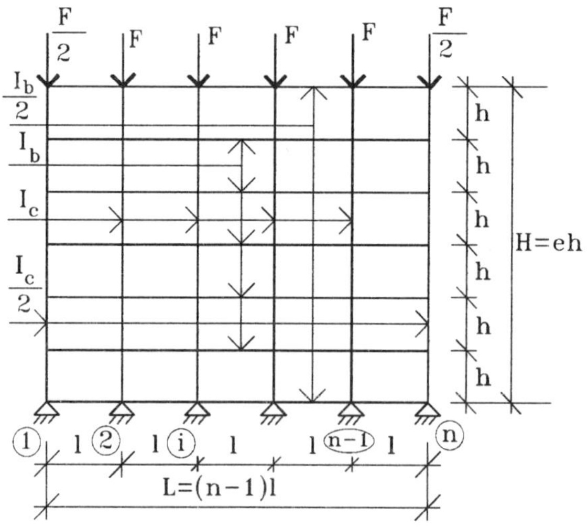

Fig. 5-18. Framework on pinned support

A framework with such characteristics is shown in Fig. 5-18.

It is emphasized again that proportionality is assumed for the correct establishment of the continuum differential equation. However, the formulae to be derived in this section also offer good approximation for regular frameworks. Frameworks on pinned supports are called regular, if all the beams and columns in the rectangular network have the same stiffness, respectively.

The continuum model will be established first, then the effect of the local and global additional buckling will be taken into consideration.

5.4.1 The Continuum Model

The substitute column is created exactly in the same way as with frameworks on fixed supports presented in Section 5.2. After cutting through the beams and putting the columns together, we arrive at the continuum model. The procedure is shown in steps in Fig. 5-19 for a proportional framework on pinned supports, subjected to concentrated forces at top floor level. The substitute column now emerges as a column on a pinned support (Fig. 5-19/e). The equilibrium of this column can only be ensured by the distributed moments along the column, representing the fixing effect of the beams.

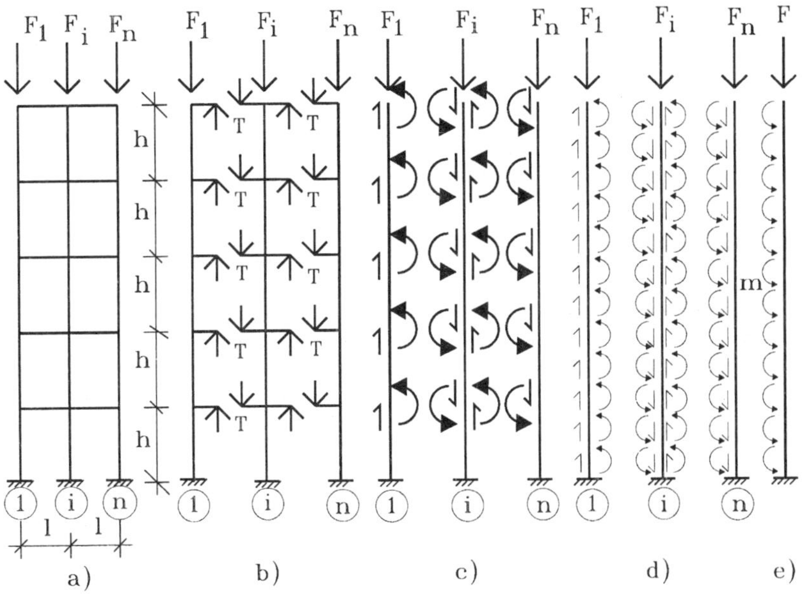

Fig. 5-19. Origination of the continuum model

5.4.2 Frameworks Subjected to Concentrated Forces at Top Floor Level

Frameworks subjected to proportional concentrated forces at top floor level will be investigated in this section. The forces are proportional if their magnitude is the same, except for the side ones whose magnitude is half of the inside ones (Fig. 5-18).

The fourth order, homogeneous differential equation of constant coefficients (5.6) derived for frameworks on fixed supports represents the governing differential equation of this substitute column as well. Because of the pinned support, however, new boundary conditions are needed. They are

$$y(0) = 0, \tag{5.51a}$$

$$y''(0) = 0, \tag{5.51b}$$

$$y''(H) = 0 \tag{5.51c}$$

and

$$y'''(0) - \frac{F - K_b}{E_c I_c} y'(0) = 0 , \tag{5.51d}$$

where the last boundary condition expresses the fact that the sum of the shear forces at the top of the substitute column equals zero. The deflection shape of the substitute column is characterized by the straight line $y = Cz$, which also satisfies the boundary conditions. Parameter C is an arbitrary constant in the equation. By making use of the above boundary conditions, we arrive at the critical load

$$F_{cr} = K_b , \tag{5.52}$$

as the closed form solution of the eigenvalue problem defined by Eqs (5.6) and (5.51a/b/c/d), where K_b is the distributed stiffness of the beams. The critical load only depends on the bending stiffness of the beams - shear stiffness from the point of view of the framework as a whole - and does not depend on the local bending stiffness of the columns. The situation was different with frameworks on fixed supports where the local critical load of the columns also appeared in the corresponding formula (see Eqn 5.11). The explanation for the lack of the local bending stiffness in formula (5.52) is simple: Because of the pinned support, the columns cannot activate their bending stiffness. Thus the substitute column only develop shear deformation.

Formula (5.52) gives the critical load of the substitute column and does not take into consideration two secondary effects existing with the actual framework. As with frameworks on fixed supports, frameworks on pinned supports also develop additional local sway between the beams and global sway involving the whole framework. These two effects will be built into the continuum model in the following.

The beams of the framework are not continuously distributed as was assumed with the continuum model and the beams exerting concentrated fixing effect on the columns only at storey levels make it possible for the columns to develop additional local sway between the beams (Fig. 5-20/a).

From the point of view of the local sway, the most unfavourable situation emerges when the stiffness of the beams is infinitely great (Fig. 5-20/b). In this case, local sway can develop at any storey level and the columns of one storey height develop buckling, with the effective length h. The corresponding critical load of a one storey unit is

$$K_c = \sum_1^n k_{c,i} = \frac{\pi^2 E_c I_c}{h^2} , \tag{5.53}$$

where I_c stands for the sum of the second moment of area of the cross-sections of the columns.

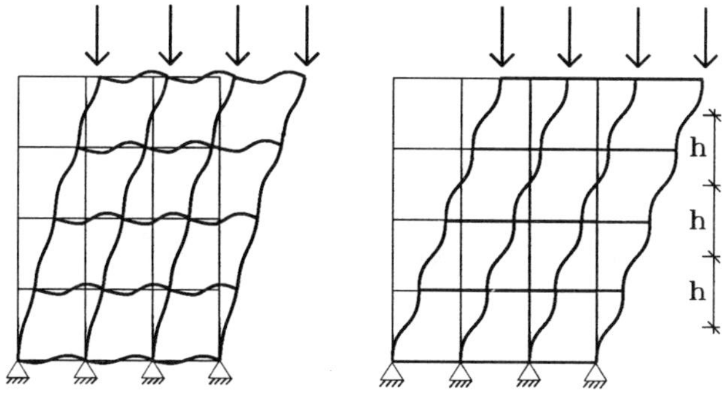

Fig. 5-20. Local sway with a) flexible and b) stiff beams

Both the critical load obtained by analyzing the continuum model (5.52) and the local critical load defined by Eqn (5.53) represent buckling failure through shear deformation. In the first case, when Eqn (5.52) was derived, it was tacitly assumed that the stiffness of the columns was great - as far as local sway was concerned -, and in the second case it was openly assumed that the stiffness of the beams was infinitely great. The two shear-type phenomena can be combined by the Föppl-Papkovics reciprocal additive theorem:

$$K = K_b \frac{K_c}{K_c + K_b} . \tag{5.54}$$

The critical load obtained from the analysis of the substitute column can now be modified to allow for the effect of the local sway. Stiffness K_b in formula (5.52) should be replaced with K:

$$F_{cr} = K = K_b \frac{K_c}{K_c + K_b} . \tag{5.55}$$

During buckling, the columns of the framework develop axial deformation. This phenomenon is similar to the buckling of a column whose longitudinal fibres undergo shortening and lengthening. The resistance of a column is characterized by the second moment of area of the cross-section. With frameworks, the longitudinal fibres are the columns of the framework and the resistance of the framework is characterized by the second moment of area of the cross-sections of the columns of the framework. We arrived at the same conclusion when frameworks on fixed supports were investigated. It is not surprising since, from the point of view of global buckling, it is all the same whether the supports are fixed or pinned. The phenomena are the same, so that the modifiers introduced for frameworks on fixed supports can be applied. On the basis of formulae

(5.22) and (5.55), we arrive at

$$F_{cr} = \frac{K}{1 + \dfrac{K}{F_g}} = \left(\frac{1}{F_g} + \frac{1}{K} \right)^{-1} . \tag{5.56}$$

In formula (5.56) K is the shear stiffness of the framework defined by (5.54) and F_g is global critical load of the framework, representing the effect of the axial deformation of the columns. Its value is obtained from formula (5.21), making use of formula (5.20).

The repeated application of Southwell's additive formula and the Föppl-Papkovics additive reciprocal formula also results in Eqn (5.56), showing that in this case the more sophisticated continuum approach leads to the same result.

5.4.3 Frameworks with UDL at Floor Levels

Frameworks on pinned supports, subjected to UDL at floor levels will be investigated in this section. The intensity of the load is $p/2$ at the top and ground floor levels and p at all other levels (Fig. 5-21/a). First, the critical load of the continuum model will be determined. Then, by introducing modifiers, the secondary effects will be taken into consideration.

The substitute column with the substitute load is produced in steps. The procedure is similar to that with frameworks on fixed supports (Fig. 5-21/a/b/c/d/e).

Equation (5.32) presented in Section 5.3.2.1 represents the governing differential equation of this substitute column. The boundary conditions complementing this equation are

$$y(0) = 0, \tag{5.57a}$$

$$y''(0) = 0, \tag{5.57b}$$

$$y''(H) = 0 \tag{5.57c}$$

and

$$y'''(0) - \frac{K_b}{E_c I_c} y'(0) = 0 , \tag{5.57d}$$

where the last condition allows for the fact that the sum of the shear forces equals zero at the top of the substitute column.

It is not easy to handle mathematically the governing differential equation with the boundary conditions and any possibility to simplify the problem should be considered. One such possibility is to reduce the order of the fourth order differential equation. By integrating the equation once and using up boundary condition (5.57d), we arrive at

$$y''' + \frac{N(z) - K_b}{E_c I_c} y' = 0. \tag{5.58}$$

The boundary conditions for this third order differential equation of variable coefficients are defined by Eqs (5.57a/b/c).

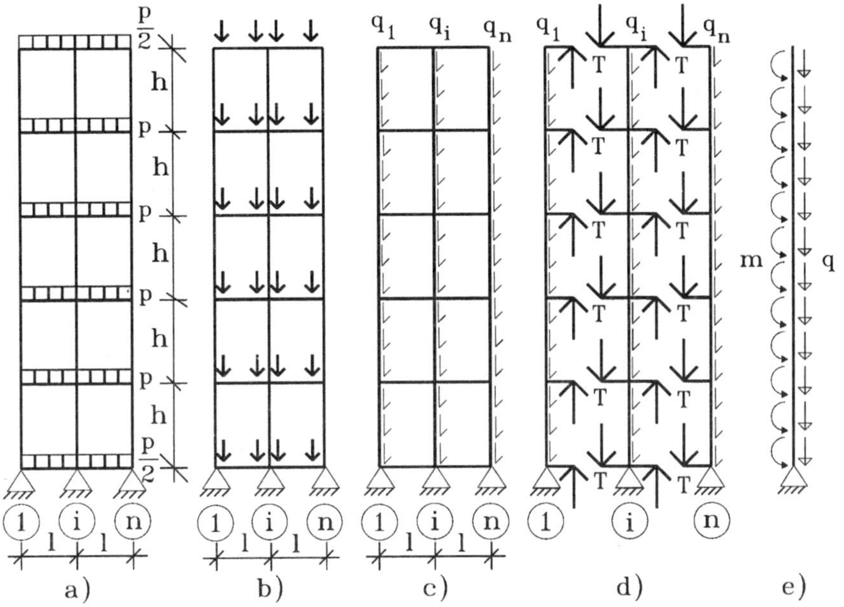

Fig. 5-21. Continuum model for frameworks under UDL

Neither Eqn (5.32) nor Eqn (5.58) with the corresponding boundary conditions has closed form solution but a lower bound can be produced by the application of Southwell's additive theorem [Chapter 8]. This procedure is not only useful in offering an initial value for the actual solution of the problem but also clearly shows how the different stiffness characteristics affect the behaviour of the substitute column.

According to Southwell's theorem, assume first that the substitute column only has bending stiffness and its shear stiffness equals zero. In this case differential equation (5.58) assumes the form

$$y''' + \frac{N(z)}{E_c I_c} y' = 0. \tag{5.59}$$

This equation together with boundary conditions (5.57a/b/c) represents a column with a free upper and a pinned lower end and with only bending stiffness. Such a column is not stable so that its critical load is

$$N_{cr} = 0. \tag{5.60}$$

Assume now that the substitute column only has shear stiffness but has no bending stiffness. Equation (5.58) assumes the form

$$y'\left(N(z) - K_b\right) = 0 \,, \tag{5.61}$$

where $N(z)=qz$, which is the uniformly distributed normal load on the substitute column. This equation is identical with Eqn (5.38), derived for the substitute column on fixed support in Section 5.3.2.1. The boundary conditions are also the same so the solution obtained there can be used:

$$N_{cr} = K_b \,. \tag{5.62}$$

The remarks on the unusual deflection of the substitute column, made in Section 5.3.2.1, still hold: When the load reaches critical value, the column develops sudden deflection at the bottom but otherwise the axis of the column remains straight. This unusual type of deflection is shown in Fig. 5-14. According to the results, columns on fixed and pinned supports, having only shear stiffness and subjected to uniformly distributed normal load behave in the same way: When the critical load is reached they develop sudden deflection but otherwise the column remains straight. In both cases the critical load equals the shear stiffness of the column.

Returning to the approximate solution, according to Southwell's theorem, the critical load of the column with *both* bending and shear stiffness is obtained by producing the sum of the part critical loads in Eqs (5.60) and (5.62):

$$N_{cr} = K_b \,. \tag{5.63}$$

Relying on this lower bound to the critical load, it is now easier to find a numerical solution to the original problem defined by differential equation (5.58) and boundary conditions (5.57a/b/c). Applying the power series method and introducing the non-dimensional quantities

$$\alpha = \frac{N_{cr}}{N_1^*} \,, \tag{5.64}$$

$$\beta = \frac{K_b}{N_1^*} \,, \tag{5.65}$$

the eigenvalue problem can be converted into a simple mathematical problem [Zalka, 1987] where the first root of a polynomial results in eigenvalue α, with which the critical load assumes the form

$$N_{cr} = \alpha N_1^* \, . \tag{5.66}$$

Parameter N_1^* in the above expressions is defined by

$$N_1^* = \frac{E_c I_c}{H^2} \, . \tag{5.67}$$

Values of α needed for the calculation of the critical load are given in Figs 5-22 and 5-23 and in Table 5-3. By making use of formula (5.65), formula (5.66) can be rearranged as

$$N_{cr} = \frac{\alpha}{\beta} K_b \, . \tag{5.68}$$

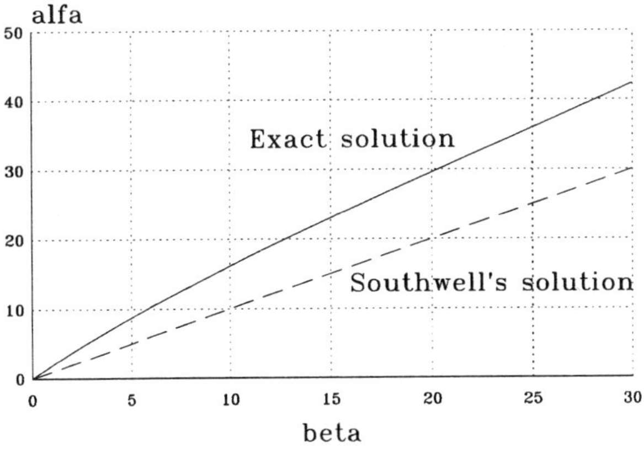

Fig. 5-22. Values for load parameter α for $0 \leq \beta \leq 30$

Formula (5-68) clearly shows the relationship between the critical load and the shear stiffness of the substitute column. It is also seen that, although the bending stiffness does not play a direct role - formula (5.60) shows that the part critical load which belongs to the bending stiffness vanishes -, the bending stiffness does have an effect on the critical load through the term α/β.

According to Table 5-3, inequality

$$\frac{\alpha}{\beta} > 1$$

holds, showing that the bending stiffness of the substitute column indirectly increases the shear stiffness.

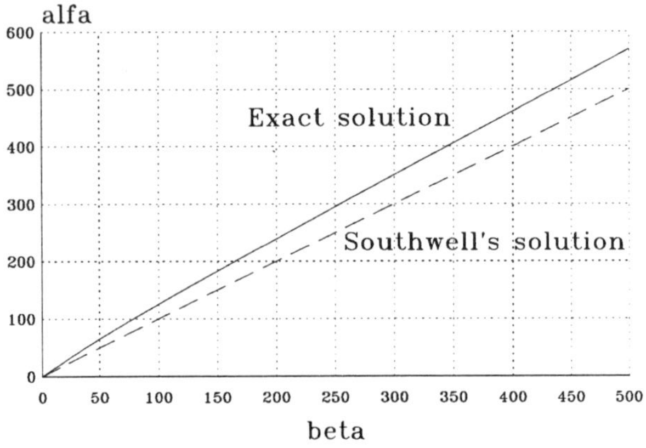

Fig. 5-23. Values for load parameter α for $0 \le \beta \le 500$

Formulae (5.63) and (5.68) make it possible to check the accuracy of Southwell's theorem. The critical load was calculated by both Southwell's additive formula (5.63) and the exact formula (5.68) for different stiffness ratios β and the deviation of the approximate value from the exact one (Δ) was determined. The comparison presented in Table 5-3 shows that Southwell's formula is always conservative. The maximum deviation is 50 per cent.

As with the frameworks subjected to concentrated forces at top floor level, the additional local sway has to be taken into consideration. However, there is some slight difference here. The normal load causing the additional sway was constant with the frameworks subjected to concentrated force and the storeys were equally vulnerable, as far as local sway was concerned. The normal load now varies along the columns and reaches its maximum value at the bottom of the columns where they have pinned supports. It follows that from the point of view of the additional local sway, it is the first storey region that has to be considered. With some approximation on the safe side, it can be assumed that the total normal load acts at the first floor level, i.e., the load increment between the ground floor level and the first floor level is also put on to the first floor level.

As formula

$$\frac{pL}{2} << epL$$

holds (Fig. 5-24), this approximation on the safe side is of little importance and its significance decreases as the number of storeys increases.

Table 5-3. Load parameter α for frameworks on pinned supports
α : continuum solution, α_s : Southwell, Δ : Southwell's error

β	α_s	α	Δ	β	α_s	α	Δ	β	α_s	α	Δ
0.000	-	-	-	0.7	0.7	1.369	49	130	130	160.0	19
0.001	0.001	0.0020	50	0.8	0.8	1.559	49	140	140	171.4	18
0.002	0.002	0.0040	50	0.9	0.9	1.749	49	150	150	182.8	18
0.003	0.003	0.0060	50	1.0	1.0	1.938	48	160	160	194.2	18
0.004	0.004	0.0080	50	2.0	2.0	3.766	47	170	170	205.5	17
0.005	0.005	0.0100	50	3.0	3.0	5.505	46	180	180	216.8	17
0.006	0.006	0.0120	50	4.0	4.0	7.172	44	190	190	228.0	17
0.007	0.007	0.0140	50	5.0	5.0	8.779	43	200	200	239.3	16
0.008	0.008	0.0160	50	6.0	6.0	10.336	42	250	250	295.2	15
0.009	0.009	0.0180	50	7.0	7.0	11.853	41	300	300	350.7	14
0.010	0.010	0.0200	50	8.0	8.0	13.334	40	350	350	405.8	14
0.020	0.020	0.0400	50	9.0	9.0	14.787	39	400	400	460.8	13
0.030	0.030	0.0599	50	10.0	10.0	16.215	38	450	450	515.5	13
0.040	0.040	0.0799	50	15.0	15.0	23.082	35	500	500	570.0	12
0.050	0.050	0.0998	50	20.0	20.0	29.656	33	600	600	678.7	12
0.060	0.060	0.1198	50	30.0	30.0	42.334	29	700	700	786.8	11
0.070	0.070	0.1397	50	40.0	40.0	54.655	27	800	800	894.6	11
0.080	0.080	0.1596	50	50.0	50.0	66.757	25	900	900	1002.0	10
0.090	0.090	0.1795	50	60.0	60.0	78.707	24	1000	1000	1109.2	10
0.100	0.100	0.1993	50	70.0	70.0	90.540	23	1200	1200	1322.8	9
0.200	0.200	0.3974	50	80.0	80.0	102.28	22	1400	1400	1535.6	9
0.300	0.300	0.5941	50	90.0	90.0	113.94	21	1600	1600	1747.8	8
0.400	0.400	0.7896	49	100.0	100.0	125.54	20	1800	1800	1959.5	8
0.500	0.500	0.9839	49	110.0	110.0	137.09	20	3000	3000	3222.2	7
0.600	0.600	1.1769	49	120.0	120.0	148.58	19	9000	9000	9001.0	0

With this approximation, we arrived at the same situation as with the frameworks subjected to concentrated forces. Again, the formula of the shear stiffness has to be modified so that it can take the local sway into consideration. By so doing, on the basis of formulae (5.54), (5.66) and (5.68), we arrive at

$$N_{cr} = \alpha N_1^* \tag{5.69}$$

or

$$N_{cr} = \frac{\alpha}{\beta} K \tag{5.70}$$

for the critical load of frameworks on pinned supports, where

$$\beta = \frac{K}{N_1^*} \tag{5.71}$$

and K and N_1^* are defined by Eqs (5.54) and (5.67), respectively. Values for α are collected in Table 5-3 and in Figs 5-22 and 5-23.

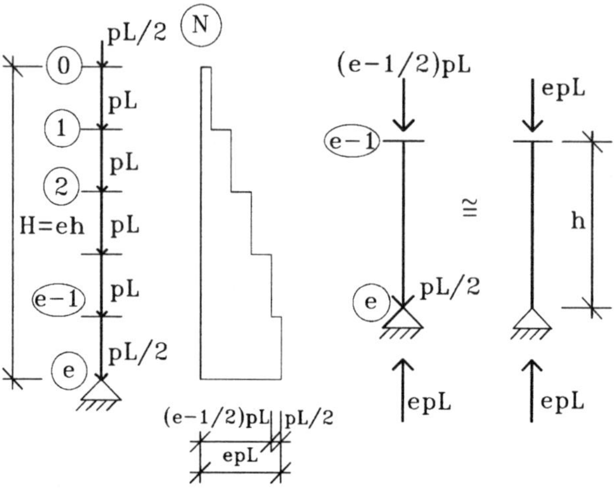

Fig. 5-24. Maximum load for the local sway

In the extreme case when the bending stiffness of the columns tends to infinity, the second term in differential equation (5.58) vanishes and consequently the equation collapses so that formulae (5.69) and (5.70) cannot be used. In this case the critical load is determined by relying on elementary statics. The equilibrium of the substitute column (Fig. 5-25) is ensured if the equality

$$N\frac{B}{2} - mH = 0$$

holds, from which we obtain

$$N_{cr} = 2K \ . \tag{5.72}$$

If the separate effects of the bending and shear stiffnesses are to be seen, then Eqn (5.69) or Eqn (5.70) should be rearranged. By so doing, the formula

$$N_{cr} = (\alpha - \beta)N_1^* + K \ , \tag{5.73}$$

is obtained, where N_1^* is defined by expression (5.67).

This interesting formula shows that even if the columns cannot activate their bending stiffness because of the pinned supports, as far as the local bending critical load is concerned, the bending stiffness still has a contribution to the resulting critical load.

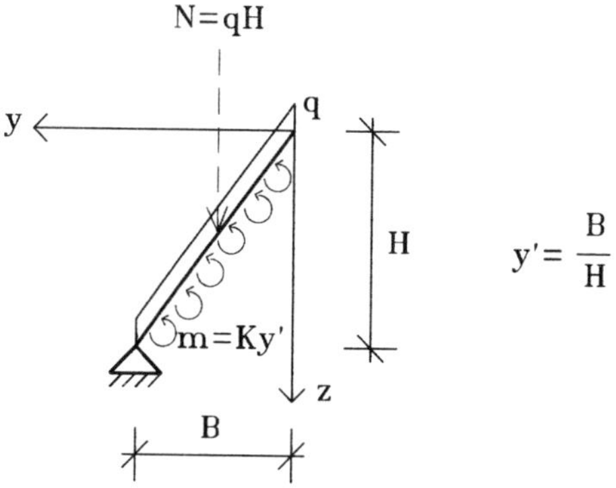

Fig. 5-25. Equilibrium with infinitely great bending stiffness

Because of the axial deformation of the columns, the formula of the critical load (5.73) should be modified. From the point of view of this effect, frameworks on fixed supports and frameworks on pinned supports behave in the same way. Due to the axial forces on the columns at floor levels, resulting from the shear forces at the contraflexure points of the beams, the columns develop axial deformations (shortening and lengthening) and global bending sway of the framework occurs. Due to the uniformly distributed nature of the load, the part critical load which belongs to this phenomenon is defined by the global critical load

$$N_g = \frac{7.837 E_c I_g}{H^2} \tag{5.74}$$

where I_g is the global second moment of area of the cross-sections of the columns, given by formula (5.20). Knowing the global critical load, formula (5.73) can be extended to allow for the effect of the additional global bending sway:

$$N_{cr} = \frac{(\alpha - \beta) N_1^* + K}{1 + \dfrac{K}{N_g}} . \tag{5.75}$$

Shear stiffness K and stiffness ratio β in formula (5.75) are defined by (5.54) and (5.71), respectively. Values for α are presented in Table 5-3 and Figs 5-22 and 5-23.

As the different types of deformation cannot develop at the same time with certain stiffness ratios, modifier r used earlier should again be introduced. By so doing, the formula

$$N_{cr} = \frac{(\alpha-\beta)rN_1^* + K}{1 + \frac{rK}{N_g}} . \tag{5.76}$$

is obtained for the critical load of frameworks on pinned supports, under UDL at floor levels.

The repeated application of Southwell's additive theorem and the Föppl-Papkovics reciprocal additive theorem now results in

$$N_{cr} = \frac{K}{1 + \frac{K}{N_g}} . \tag{5.77}$$

According to the nature of the additive theorems, formula (5.77) always gives a conservative critical load. The main difference between formulae (5.76) and (5.77) is in their numerator. The numerator represents the exact solution in (5.76), and Southwell's approximate solution in (5.77), when the effect of the axial deformations is neglected. The accuracy of Southwell's theorem was already investigated in this Section: It is always conservative with the maximum error of 50 per cent.

5.5 LOADING

It was the regularity of the structural systems, which made it possible to establish simple equations and to arrive at simple formulae for the critical load. It is not enough that the geometry and the stiffness characteristics of the structure are regular. Regularity in the loading is also required. Two typical cases were assumed in the previous sections. The load case when the structure is only subjected to concentrated forces at the top end of the columns does not represent a common practical case. As a rule, multistorey structures carry uniformly distributed loads (UDL) on the beams. The intensity of this UDL is often the same at each floors. On the top level, however, there are sometimes more structural layers or some other heavier objects, resulting in slightly more load. This "surplus" load can be considered as a concentrated load at the top of the building. In this case, we have a combination of UDL at floor levels and concentrated forces at top floor level.

5.5.1 Structures with Concentrated Loads and UDL

Dunkerley's reciprocal additive theorem [Chapter 8] makes it possible to investigate structures subjected to more than one system of load at the same time. Applying the reciprocal formula

$$\frac{F}{F_{cr}} + \frac{N}{N_{cr}} \leq 1 \tag{5.78}$$

to frameworks subjected to *both* concentrated forces at top floor level and uniformly distributed loads at floor levels, we can take into consideration the simultaneous effect of the two load systems. In formula (5.78) F_{cr} is the critical load of the structure subjected to concentrated forces on top floor level, N_{cr} is the critical load of the structure subjected to uniformly distributed load at floor levels, F is the sum of the applied concentrated forces at the top of the structure and N is the sum of the applied load of the structure subjected to uniformly distributed load at floor levels, defined by formula (5.31).

With exact critical loads, formula (5.78) offers a conservative estimate and can be rearranged to show the safety of a structure against buckling due to the two simultaneous load systems:

$$\nu = \frac{1}{\dfrac{F}{F_{cr}} + \dfrac{N}{N_{cr}}} . \tag{5.79}$$

In formula (5.79) ν is the load factor giving the ratio of the overall critical load of the structure to the applied load.

5.5.2 Correction for Frameworks Subjected to UDL

When frameworks on fixed supports, subjected to UDL at floor levels were investigated, it was pointed out that the distribution of the load acting at floor levels *downwards* (Fig. 5-26/a/b) resulted in an unfavourable approximation. By making use of Dunkerley's theorem, the effect of this approximation can be reduced. In fact, by the application of the formula

$$N_{cr}^{D} = \left(\frac{r}{2eF_{cr}} + \frac{1}{N_{cr}}\right)^{-1} , \tag{5.80}$$

the approximation is made on the safe side. In formula (5.80) e is the number of storeys and N_{cr}^{D} allows for the fact that by distributing the loads downwards, a "surplus" concentrated force of magnitude $N/2e$ results from the procedure (Fig. 5-26/c). The effect of this concentrated force may be significant for low-rise structures and can be negligible for tall structures.

The introduction of reduction factor r defined by formula (5.23) into formula (6.6) controls the effect of Dunkerley's correction. When the value of r decreases, the effect of Dunkerley's correction also decreases. As the numerical examples in Section 5.6.1 show, the buckling shape of the columns of the framework with small r tends to a straight vertical line above the first storey region. The distribution and the location of the applied load is no longer important in such cases and reduction factor r in formula (5.80) reduces the effect of Dunkerley's correction accordingly.

To illustrate the order of magnitude of this effect, the critical load of the two-bay, four-storey framework subjected to uniformly distributed load at floor levels, shown in Fig. 5-27, was computed. Formula (5.50) gave

$$N_{cr} = 346.06 \text{ MN}$$

and formula (5.80) resulted in

$$N_{cr}^{D} = 278.80 \text{ MN}.$$

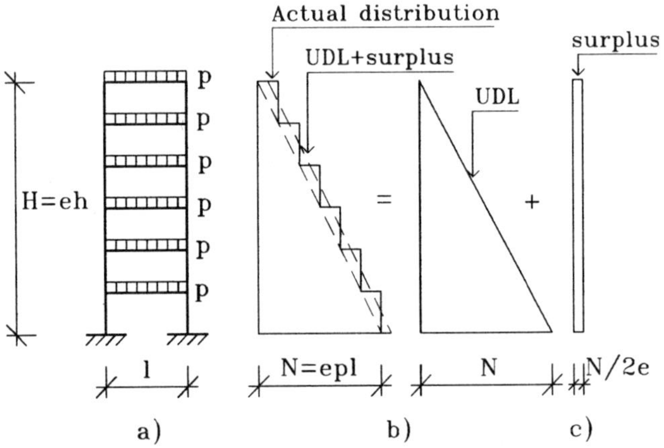

Fig. 5-26. Surplus load at top level with frameworks under UDL

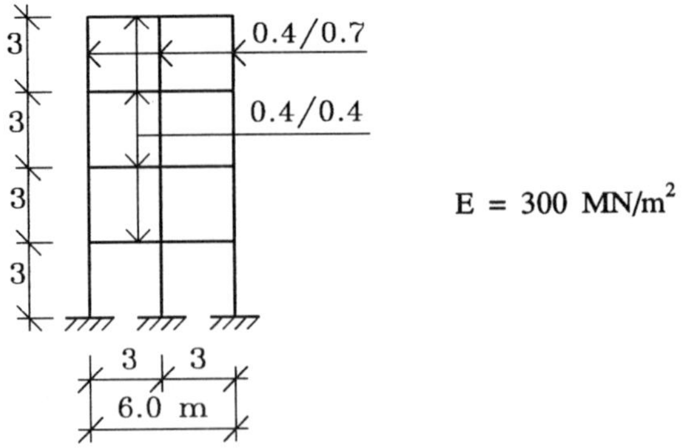

Fig. 5-27. Four-storey framework subjected to UDL

The exact critical load is

$$N_{cr} = 305.16 \text{ MN} .$$

The critical load obtained from formula (5.50) is 13.40 per cent *greater* than the exact value while formula (5.80) resulted in a conservative estimate, a critical load 8.66 per cent smaller than the exact one. A detailed accuracy analysis is presented in Section 5.6.

5.5.3 The Effect of Non-proportional Loading

The formulae obtained in the foregoing are applicable to obtaining the *total* critical load. The *total* critical load is of interest since one column cannot fail alone but a whole storey with all the columns [Dulácska and Kollár, 1960; Armer and Grimer, 1972]. Even if one or more columns are overloaded, the framework as a whole can remain stable. This is only possible if the stronger - or underloaded - columns support the weaker - or overloaded - columns. It follows that there is some load redistribution. However, the non-proportional loading of the framework does have an effect on the accuracy of the formulae obtained by the continuum method. The less proportional the load of the structure, the less reliable the results are. The question is, how much. According to Dulácska's theorem, the critical load of a single storey framework hardly depends on the distribution of the external vertical load among the columns [Dulácska and Kollár, 1960]. He suggests that a maximum of four-fold overloading should be allowed in which case the error made by calculating the *total* critical load is less than 10 per cent. The formula he suggests assumes the form

$$\frac{F_i}{F^*_{i,cr}} \leq 4 \frac{\Sigma F_i}{\Sigma F^*_{i,cr}} \qquad (5.81)$$

where

$$F^*_{i,cr} = \frac{\pi^2 EI_i}{h^2_{o,i}}$$

is the critical load of the *i*-th column,

$$h_{o,i} = \nu h_i$$

is the effective length of the *i*-th column and F_i is the load on the *i*-th column.

Numerical data on the effect of the non-proportional loading is presented in Section 5.6.

5.6 ACCURACY ANALYSIS, CHARACTERISTIC DEFORMATIONS AND DESIGN EXAMPLES

The derivation of the formulae of the critical load was based on the assumption that the frameworks were proportional structures. Proportionality with frameworks on fixed supports means that all the columns have the same stiffness, except for the side ones whose stiffness is half of the others and all the beams have the same stiffness, except for those at top level whose stiffness is half of the others. It is a common practical case, however, that all the columns have the same stiffness and also all the beams have the same stiffness. The formulae derived can still be used, with two remarks. Structures with identical beams are slightly stiffer than the proportional ones so the critical load obtained by the continuum method is slightly conservative. The situation with identical columns is somewhat different. If all the columns have the same stiffness, then the assumption that all the columns develop the same deflection is not satisfied since the "uniform" stiffness of the bays is disturbed in the side bays. The difference is not significant and the error resulting from this discrepancy is less than 5 per cent [Kollár, 1972]. Chapter 6 deals with different possibilities of the generalization of the continuum method.

The usefulness of any approximate procedure is proportional to its accuracy. In order to establish this important feature, an accuracy analysis was carried out [Zalka, 1991], involving hundreds of frameworks and the results of a small-scale test. The results of a sophisticated finite element procedure, considered to be "exact" were compared to the results of the approximate formulae presented in this Section. The accuracy of Southwell's additive formulae and that of the Föppl-Papkovics reciprocal additive formula was also evaluated. This numerical analysis also made it possible to show to what extent the different stiffness characteristics effect the value of the critical load. A summary of this numerical analysis is given bellow.

The structures for the numerical analysis are subjected to both concentrated forces at the top of the columns and UDL on the beams. The size of the frameworks varies from four to twenty five storeys and from one bay to four bays, with different stiffness characteristics. The length of the beams and columns is 3 m. The cross-section of the individual elements is given in the corresponding figures. The value of the modulus of elasticity is E=3x10^4 MN/m^2. The buckling shape of the frameworks is given in the figures. The nature of the shapes is emphasized, local double curvature bending between the nodal points is only shown if it is the characteristic mode of failure.

Four sets of structure are investigated. They cover two practical cases: Frameworks with columns whose stiffness is slightly greater than that of the beams and frameworks with beams whose stiffness is greater than that of the columns. The other two sets are extreme, theoretical cases. Frameworks with columns whose stiffness is by orders of magnitude greater than that of the beams represent the real "bending-type" category. Finally, frameworks with practically infinite beam stiffness fall into the real "shear-type" category. The inclusion of the extreme cases makes it possible to make a well-based judgment on the accuracy of the approximate formulae.

The critical loads computed by the exact method, the continuum method and the additive formulae for concentrated load at the top of the columns and UDL are presented in tables. The tables also give the error in per cents, made by the application of the approximate formulae. Positive error means conservative estimate (the approximate critical load is smaller than the exact one) and the error is marked negative when the approximate critical load is greater than the exact one.

The first structure, framework FF1 (Fig. 5-28), is used to show in detail how the actual computation of the critical load is carried out in the structural design process.

5.6.1 Frameworks on Fixed Supports

The critical load of frameworks on fixed supports subjected to concentrated forces at the top of the columns is computed from Eqn (5.24) for the continuum analysis and from Eqn (5.22) when the accuracy of the additive formulae is compared to the exact critical load. When the frameworks are subjected to UDL on the beams, Eqn (5.80) is used for the continuum analysis and Eqn (5.49) is used as the solution which belongs to the repeated application of the additive theorems.

With the first structure, framework FF1 shown in Fig. 5-28, the computation of the sixteen-storey case is presented in detail.

The two-bay framework is subjected to three concentrated forces at the top of the columns (Fig. 5-28/a) and UDL on the beams (Fig. 5-28/c). The framework is non-proportional, whose columns and beams are identical, respectively. The stiffness of the columns of the framework is greater than that of the beams.

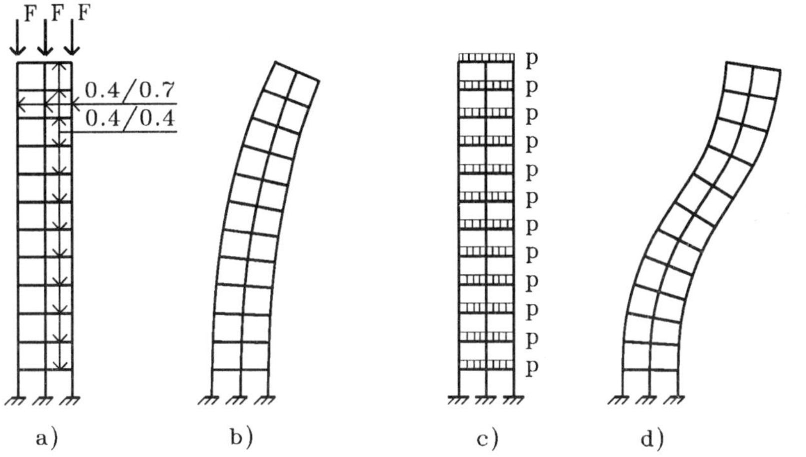

Fig. 5-28. Framework FF1

The following stiffness characteristics are needed for the computation.

The stiffness of the beams is defined by formulae (5.9) and (5.5):

$$K_b = 2\sum_1^{n-1} k_{b,i} = 4\frac{6E_bI_b}{lh} = \frac{24\text{x}30{,}000\text{x}0.002133}{3\text{x}3} = 170.67 \text{ MN}.$$

The stiffness of the columns is given by formula (5.14):

$$K_c = \sum_1^n k_{c,i} = \sum_1^n \frac{\pi^2 E_c I_{c,i}}{h^2} = \frac{3\text{x}\pi^2\text{x}30{,}000\text{x}0.011433}{3^2} = 1128.42 \text{ MN}.$$

The shear stiffness is obtained from formula (5.16):

$$K = K_b \frac{K_c}{K_c + K_b} = 170.67 \frac{1128.42}{1128.42 + 170.67} = 148.25 \text{ MN},$$

where the reduction factor (5.23), characteristic of the behaviour of the framework, assumes the value

$$r = \frac{K_c}{K_c + K_b} = 0.87.$$

The local bending critical load is obtained from formula (5.13) for the concentrated load:

$$F_l = \frac{\pi^2 E_c I_c}{4H^2} = \frac{\pi^2\text{x}30{,}000\text{x}3\text{x}0.011433}{4\text{x}48^2} = 1.1 \text{ MN},$$

and from formula (5.37) for the UDL:

$$N_l = \frac{7.837 E_c I_c}{H^2} = \frac{7.837\text{x}30{,}000\text{x}3\text{x}0.011433}{48^2} = 3.50 \text{ MN},$$

The global bending critical load assumes the value

$$F_g = \frac{\pi^2 E_c I_g}{4H^2} = \frac{\pi^2\text{x}30{,}000\text{x}5.04}{4\text{x}48^2} = 161.92 \text{ MN}$$

for concentrated forces [formula (5.21)] and

$$N_g = \frac{7.837 E_c I_g}{H^2} = \frac{7.837\text{x}30{,}000\text{x}5.04}{48^2} = 514.30 \text{ MN},$$

when the framework is under UDL [formula (5.47)]. The global second moment of area in the above formulae is obtained from formula (5.20):

$$I_g = \sum_1^n A_{c,i} t_i^2 = 2x0.4x0.7x3^2 = 5.04 \text{ m}^4.$$

Stiffness ratio β is also needed for the computation (formula 5.45):

$$\beta = \frac{K}{N_1} = \frac{148.25}{3.50} = 42.35.$$

The load parameter is now obtained from Table 5-2 in the function of β as

$$\alpha = 60.49.$$

With the above stiffness and geometrical characteristics, the critical loads can now be produced.

The critical load for frameworks under concentrated forces is obtained from formula (5.24):

$$F_{cr} = \frac{0.87\times1.1 + 148.25 + \dfrac{0.87\times1.1\times148.25}{161.92}}{1 + \dfrac{148.25}{161.92}} = 78.35 \text{ MN}.$$

If the framework is subjected to UDL on the beams, the critical load is obtained in two steps. Neglecting the effect of the excess load at the top, the critical load is obtained from Eqn (5.50):

$$N_{cr} = \frac{0.87\times3.50(60.49\text{-}42.35) + 148.25 + \dfrac{0.87\times3.5\times148.25}{514.30}}{1 + \dfrac{0.87\times148.25}{514.30}} = 163.36 \text{ MN}$$

In making use of Eqs (5.24) and (5.50), Dunkerley's additive formula (5.80) takes the excess load at top into consideration:

$$N^D_{cr} = \left(\frac{0.87}{2\times16\times78.35} + \frac{1}{163.39}\right)^{-1} = 154.62 \text{ MN}.$$

Assume that the concentrated forces on the columns assume the value

$$F_i = 1 \text{ MN}$$

respectively and the intensity of the UDL on the floor levels is

$$p = 30 \text{ kN/m}.$$

In this case the total external concentrated load is

$$F = \Sigma F_i = 3 \text{ MN}.$$

The total UDL is obtained from formula (5.31):

$$N = pLe = 0.03\text{x}6\text{x}16 = 2.88 \text{ MN}.$$

With the external total loads and the critical loads now available, the ratio of the overall critical load to the applied load is obtained from formula (5.79):

$$\nu = \frac{1}{\dfrac{F}{F_{cr}} + \dfrac{N}{N_{cr}}} = \frac{1}{\dfrac{3}{78.35} + \dfrac{2.88}{154.62}} = 17.57$$

According to this result, the framework can withstand 17.57 times as much load as the currently applied load, as far as in-plane stability is concerned. For the actual structural design, however, the application of safety factors in accordance with the relevant code of practice may have to be considered.

Relying on the Föppl-Papkovics additive formulae, the critical loads are obtained from formulae (5.22) and (5.49), respectively. The approximate critical load for the concentrated load assumes the value:

$$F_{cr} = F_l + \left(\frac{1}{F_g} + \frac{1}{K}\right)^{-1} = 78.5 \text{ MN}$$

Table 5-4. Critical loads for framework FF1

No.	Critical concentrated load [MN]					Critical UDL [MN]				
of storeys	Exact method	Additive method	Δ %	Continuum method	Δ %	Exact method	Additive method	Δ %	Continuum method	Δ %
4	177.5	157.9	11	155.6	12	305.3	201.6	34	278.8	9
8	131.8	125.0	5	124.5	6	226.7	152.3	33	214.1	6
12	103.0	99.8	3	99.6	3	192.0	133.8	30	179.9	6
16	80.1	78.5	2	78.4	2	166.8	118.6	29	154.6	7
20	62.6	61.7	1	61.6	1	143.6	104.5	27	133.6	7
24	49.4	48.9	1	48.9	1	122.0	91.5	25	115.6	5

When the framework is under UDL, the the Föppl-Papkovics approximate critical load is

$$N_{cr} = \left(\frac{1}{N_g} + \frac{1}{K} \right)^{-1} = 118.6 \text{ MN.}$$

The critical loads computed above, the exact critical loads obtained by using a sophisticated FE procedure and the errors made by the application of the approximate formulae are given in Table 5-4.

Figs 5-28/b and 5-28/d shows the buckling shape at failure. According to the numerical results and to the buckling shapes, global bending buckling and global shear play an important role with bending buckling dominating with the concentrated load. The situation is similar with the UDL but here global shear is dominant.

In the following examples all the critical loads are given in the corresponding tables, without presenting the details of the computation.

Figure 5-29 shows the non-proportional, four-bay framework FF2. All the columns and the beams are identical, respectively. The stiffness of the beams is greater than that of the columns. The characteristic ratio assumes the value

$$r = \frac{K_c}{K_c + K_b} = \frac{351}{351 + 1829} = 0.16 \ .$$

The framework is subjected to five identical concentrated forces on the columns (Fig. 5-29/a) and UDL of the same intensity on every beam (Fig. 5-29/c). When the structure is subjected to concentrated forces, the combination of global bending and global shear deformation determines the buckling shape (Fig. 5-29/b). The framework under UDL develops shear deformation (Fig. 5-29/d).

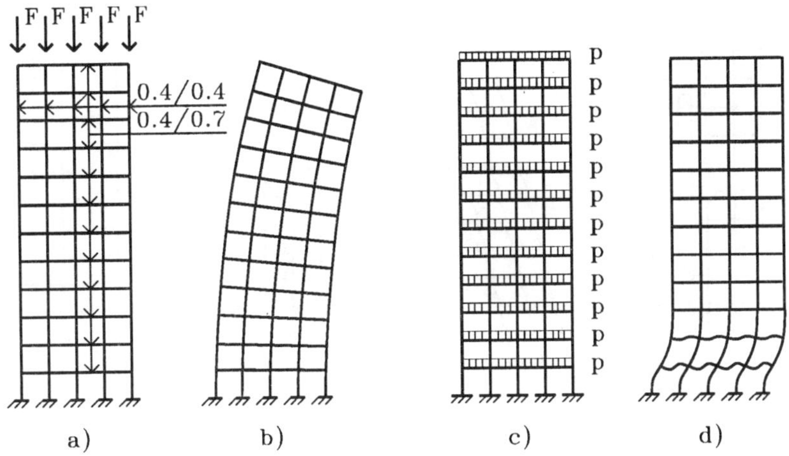

Fig. 5-29. Framework FF2

Table 5-5 contains the values of the critical loads and the errors made by the application of the approximate procedures.

Table 5-5. Critical loads for framework FF2

No.	Critical concentrated load [MN]					Critical UDL [MN]				
of storeys	Exact method	Additive method	Δ %	Continuum method	Δ %	Exact method	Additive method	Δ %	Continuum method	Δ %
4	285.8	288.7	-1	284.1	1	329.9	308.2	7	316.7	4
8	252.4	255.4	-1	254.2	-1	324.8	284.8	12	305.0	6
12	216.0	217.4	-1	216.9	-0	318.5	266.6	16	298.4	6
16	181.0	180.3	0	180.0	1	310.7	246.4	21	292.3	6
20	149.1	147.9	1	147.7	1	297.4	224.9	24	286.0	4

Framework FF5 in Fig. 5-30 has identical columns and identical beams, with extreme stiffness characteristics. The stiffness of the columns is greater by orders of magnitude than that of the beams, making the reduction factor nearly vanish:

$$r = \frac{K_c}{K_c + K_b} = \frac{5.48}{5.48 + 341333} = 0.000016 \ .$$

The load system consists of five identical forces on the columns (Fig. 5-30/a) and UDL on each beam (Fig. 5-30/c).

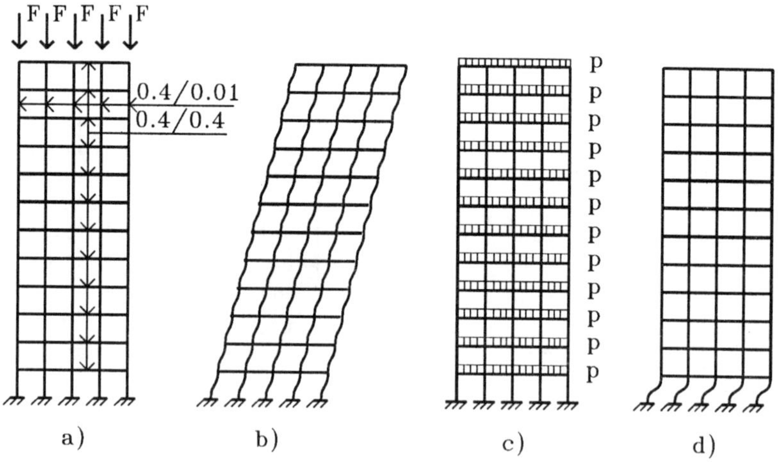

Fig. 5-30. Framework FF5

Because of the weak columns, local shear deformation dominates the buckling shape in both load cases (Figs 5-30/b and 5-30/d). The critical loads and the errors are presented in Table 5-6.

Table 5-6. Critical loads for framework FF5

No. of storeys	Critical concentrated load [kN] Exact method	Additive method	Δ %	Continuum method	Δ %	Critical UDL [kN] Exact method	Additive method	Δ %	Continuum method	Δ %
4	5.470	5.569	-2	5.483	-0	5.474	5.755	-5	5.483	-0
8	5.465	5.504	-1	5.482	-0	5.474	5.551	-1	5.483	-0
12	5.455	5.491	-1	5.482	-0	5.474	5.513	-1	5.483	-0
16	5.450	5.486	-1	5.480	-1	5.474	5.499	-0	5.483	-0
20	5.445	5.482	-1	5.479	-1	5.474	5.493	-0	5.483	-0

Another extreme case is seen in Fig. 5-31. Framework FF7 has the same geometrical arrangement and load system as the previous framework but this time the stiffness of the columns is greater by orders of magnitude than that of the beams. The reduction factor assumes the value

$$r = \frac{K_c}{K_c + K_b} = \frac{350919}{350919 + 5.33} = 0.999985 \; .$$

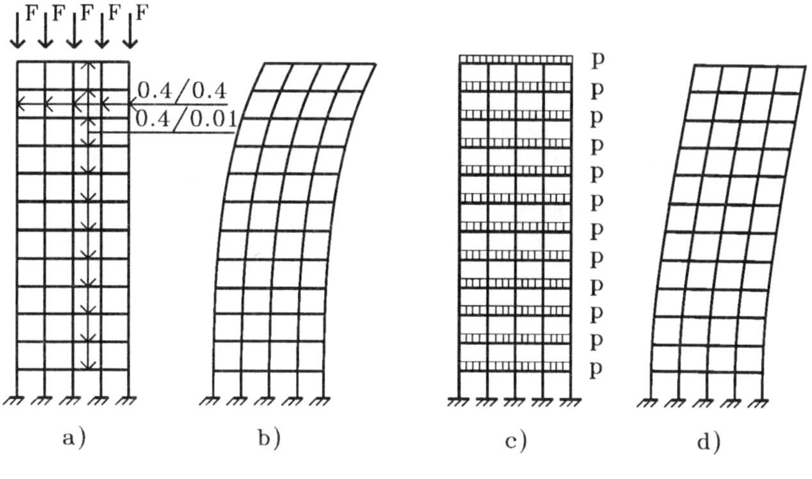

Fig. 5-31. Framework FF7

The framework now behaves like a cantilever (Figs 5-31/b and 5-31/d) and the buckling shape is characterized by the bending of the individual columns - local bending deformation. Table 5-7 gives the critical loads and the error of the approximate methods.

Table 5-7. Critical loads for framework FF7

No.	Critical concentrated load [MN]					Critical UDL [MN]				
of storeys	Exact method	Additive method	Δ %	Continuum method	Δ %	Exact method	Additive method	Δ %	Continuum method	Δ %
4	5.381	5.489	-2	5.488	-2	11.734	17.421	-48	12.478	-6
8	1.376	1.376	0	1.376	0	3.592	4.359	-21	3.646	-2
12	0.616	0.615	0	0.615	0	1.725	1.940	-12	1.723	0
16	0.349	0.348	0	0.348	0	1.007	1.094	-9	1.005	0
20	0.225	0.225	0	0.225	0	0.662	0.702	-6	0.660	0

5.6.2 Frameworks on Pinned Supports

The two types of structure considered here are only different in whether or not they have a set of beams on ground floor level. The continuum approach is very similar in the two cases. Frameworks with ground floor beams, as the basic case, are dealt with in Section 5.4 in detail. The necessary modification for frameworks without ground floor beams is given later, in Section 6.1.2. The reason why both cases are included here is that in this way a comprehensive accuracy analysis is presented and it is possible to see and compare the different structures in one place.

5.6.2.1. Frameworks with Ground Floor Beams

The critical load of frameworks on pinned supports, having ground floor beams, subjected to concentrated forces at the top of the columns is computed from Eqn (5.56) for both the continuum analysis and also when the accuracy of the additive formulae is compared to the exact critical load since the two solutions coincide. When the frameworks are subjected to UDL on the beams, Eqn (5.76) is used for the continuum analysis and Eqn (5.77) is used when the accuracy of the additive formulae is investigated.

Framework PFG3 in Fig. 5-32 has identical columns. Because of this, the structure is non-proportional. The stiffness of the columns is greater than that of the beams. All the beams have the same stiffness, except for the beams at the top and ground floor levels, whose stiffness is half of the inside ones. In this way, the distribution of the concentrated fixing effect of the beams results in a continuous flow of couples of constant intensity. The structure is subjected to three identical concentrated forces on the columns (Fig. 5-32/a) and UDL on the beams (Fig. 5-32/c).

The characteristic stiffness ratio assumes the value

$$r = \frac{K_c}{K_c + K_b} = \frac{1128}{1128 + 171} = 0.87 \ .$$

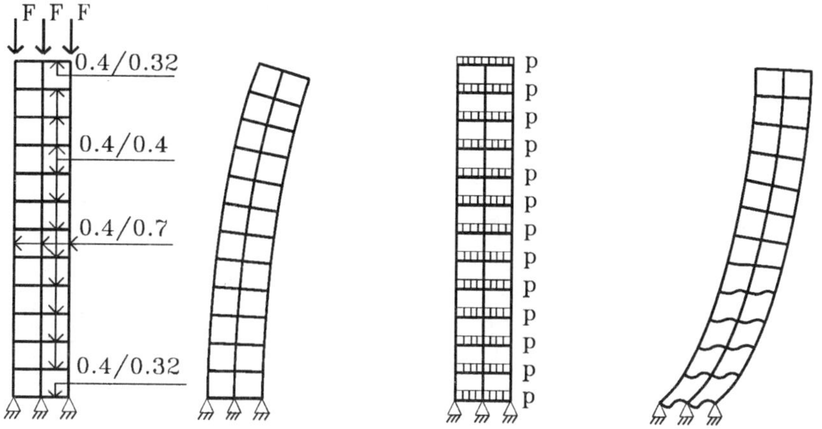

a) b) c) d)

Fig. 5-32. Framework PFG3

The buckling shape of the framework is determined by the global bending and the global shear deformation. When the structure is subjected to concentrated forces at the top of the columns, global bending is dominant (Fig. 5-32/b). With UDL on the floors, global shear plays the most important role (Fig. 5-32/d).

Table 5-8. Critical loads for framework PFG3

No. of storeys	Critical concentrated load [MN]					Critical UDL [MN]				
	Exact method	Additive method	Δ %	Continuum method	Δ %	Exact method	Additive method	Δ %	Continuum method	Δ %
4	142.68	140.22	2	140.22	2	220.32	145.62	34	206.31	6
8	123.60	120.63	2	120.63	2	185.28	138.28	25	172.79	7
12	99.84	97.85	2	97.85	2	170.69	127.56	25	152.68	11
16	78.66	77.39	2	77.39	2	159.36	115.08	28	135.32	15
20	61.80	60.99	1	60.99	1	144.67	102.21	29	119.34	18
24	49.02	48.45	1	48.45	1	125.57	89.93	28	104.80	17

The critical loads for both loading cases, computed by the exact method, the continuum method and the additive formulae are given in Table

5-8. Table 5-8 also presents the error made by the application of the approximate formulae.

The four-bay framework PFG4 shown in Fig. 5-33 is proportional. The stiffness of the columns is the same, except for the side ones whose stiffness is half of the others. The situation is similar with the beams: All have the same stiffness but for those at the top and ground floor levels, whose stiffness is half of the others. The framework is subjected to five concentrated forces of identical magnitude at the top of the columns and UDL on the beams. Neither the concentrated nor the UD load is proportional.

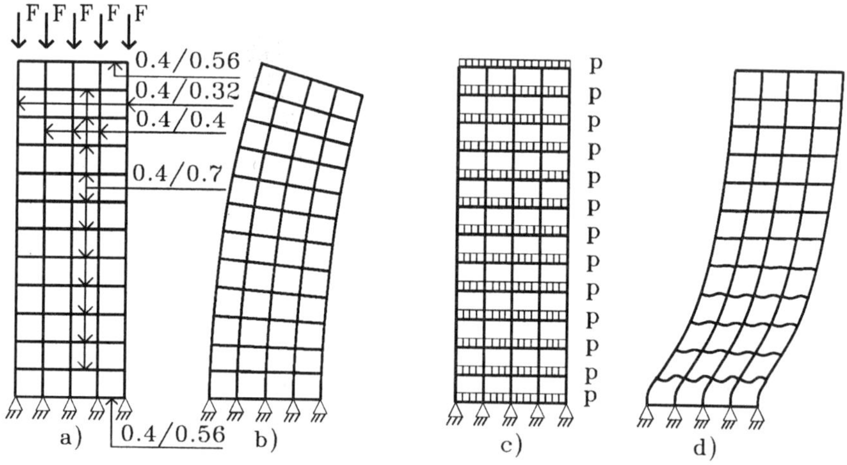

Fig. 5-33. Proportional framework PFG4

Table 5-9. Critical loads for framework PFG4

No. of storeys	Critical concentrated load [MN]					Critical UDL [MN]				
	Exact method	Additive method	Δ %	Continuum method	Δ %	Exact method	Additive method	Δ %	Continuum method	Δ %
4	239.4	234.2	2	234.2	2	291.3	240.4	17	250.3	14
8	215.5	210.3	2	210.3	2	270.9	231.9	14	246.1	9
12	185.4	179.7	3	179.7	3	263.9	219.0	17	243.0	8
16	154.9	149.3	4	149.3	4	260.0	203.1	22	239.7	8
20	127.5	122.6	4	122.7	4	256.0	185.8	27	235.9	8

The characteristic ratio assumes the value

$$r = \frac{K_c}{K_c + K_b} = \frac{281}{281 + 1829} = 0.13 ,$$

showing that the stiffness of the beams is greater than that of the columns. Global bending deformation combined with global shear deformation is characteristic of the failure shape when the structure is subjected to concentrated forces at top level (Fig. 5-33/b). With UDL on the beams, local shear deformation at ground floor level causes failure. This local shear deformation is combined with global shear deformation (Fig. 5-33/d). The critical loads and the corresponding error are tabulated in Table 5-9.

Framework PFG6 in Fig. 5-34 has extreme stiffness characteristics. The stiffness of all the beams is the same. This goes for the columns as well but their stiffness is by orders of magnitude smaller.

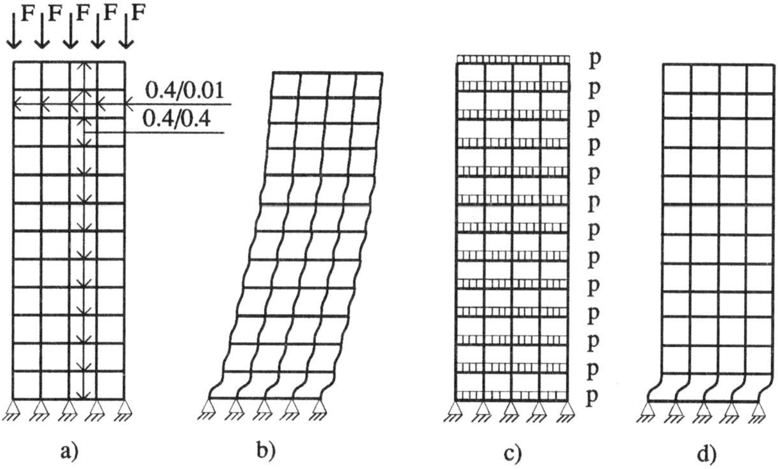

Fig. 5-34. Framework PFG6

The characteristic ratio assumes the value

$$r = \frac{K_c}{K_c + K_b} = \frac{5.48}{5.48 + 341333} = 0.000016 \ .$$

Because of the weak columns, the buckling shape is solely determined by the local shear deformation. With concentrated forces at the top of the columns, all storeys undergo local shear-type sway (Fig. 5-34/b). When the load is UDL, only the first storey develops sway since it it there where the vertical load is maximum (Fig. 5-34/d).

The critical loads and the error of the approximate methods are presented in Table 5-10.

The geometry and the two load systems of framework PFG8 in Fig. 5-35 are identical to framework PFG6 in the previous case. The stiffness characteristics are also extreme but this time it is the columns whose stiffness is by orders of magnitude greater.

Table 5-10. Critical loads for framework PFG6

No. of storeys	Critical concentrated load [kN] Exact method	Additive method	Δ %	Continuum method	Δ %	Critical UDL [kN] Exact method	Additive method	Δ %	Continuum method	Δ %
4	5.430	5.483	-1	5.483	-1	5.483	5.483	0	5.483	0
8	5.400	5.482	-1	5.482	-1	5.483	5.483	0	5.483	0
12	5.375	5.482	-2	5.482	-2	5.483	5.483	0	5.483	0
16	5.355	5.480	-2	5.480	-2	5.483	5.483	0	5.483	0
20	5.340	5.479	-3	5.479	-3	5.483	5.482	0	5.483	0

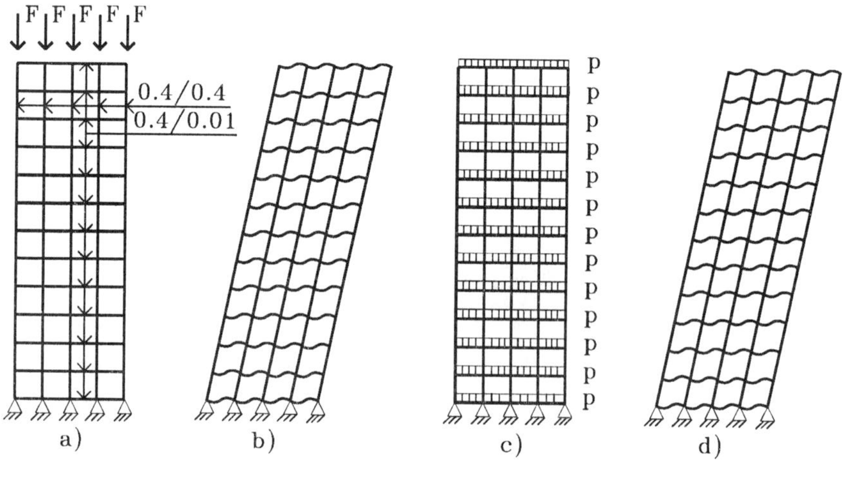

Fig. 5-35. Framework PFG8

The characteristic stiffness ratio assumes the value

$$r = \frac{K_c}{K_c + K_b} = \frac{350919}{350919 + 5.33} = 0.999985 \ .$$

Due to the pinned supports and the stiff columns, the deformation at failure is characterized by a straight line in both load cases (Figs 5-35/b and 5-35/d). The critical loads and the error of the approximate procedures are given in Table 5-11. The exact critical loads in Table 5-11 were obtained by relying on elementary equilibrium considerations since the available FE program could not cope with the problem. It was assumed for this calculation, that the stiffness of the upmost and the ground floor beams was half of the others.

Table 5-11. Critical loads for framework PFG8

No.	Critical concentrated load [kN]					Critical UDL [kN]				
of storeys	Exact method	Additive method	Δ %	Continuum method	Δ %	Exact method	Additive method	Δ %	Continuum method	Δ %
4	5.333	5.333	0	5.333	0	10.666	5.333	50	10.666	0
8	5.333	5.333	0	5.333	0	10.666	5.333	50	10.663	0
12	5.333	5.333	0	5.333	0	10.666	5.333	50	10.659	0
16	5.333	5.333	0	5.333	0	10.666	5.333	50	10.653	0
20	5.333	5.333	0	5.333	0	10.666	5.333	50	10.645	0

5.6.2.2. Frameworks without Ground Floor Beams

Section 6.1.2 shows that the continuum method can be generalized and different non-proportional and even non-regular frameworks can be investigated. We have one such case when the method is applied to frameworks on pinned supports, with missing ground floor beams. The formulae derived in Section 5.4 and used in the previous Section for the accuracy analysis of frameworks with ground floor beams can still be used with one exception. Instead of formula (5.53), formulae (6.2) and (6.3) should be used for taking into account the effect of the local shear deformation.

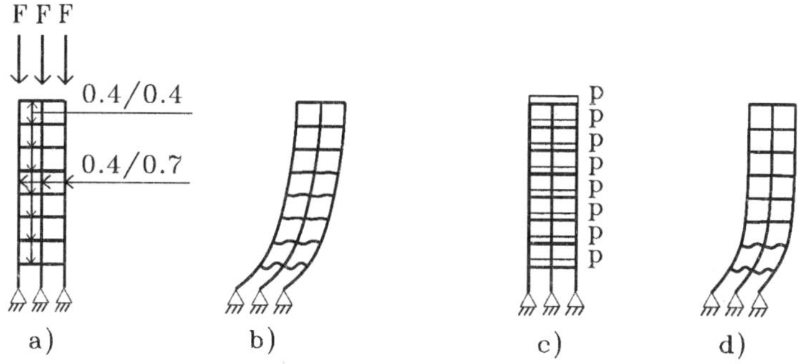

Fig. 5-36. Eight-storey framework PF1

Framework PF1 in Fig. 5-36 is a two-bay, non-proportional framework with identical columns and beams, respectively. The stiffness of the columns is greater than that of the beams. The characteristic ratio assumes the value

$$r = \frac{K_c}{K_c + K_b} = \frac{282}{282 + 171} = 0.623 \; .$$

The buckling shape of the framework is interesting. For the first time, the nature of the shape depends on the height of the framework. With low-rise frameworks subjected to either concentrated forces at top level or UDL on the beams, combined local and global shear deformation is characteristic when a great part of the deformation develops in the first storey region where, because of the pinned supports, the framework is most vulnerable to local sway (Figs 5-36/b and 5-36/d).

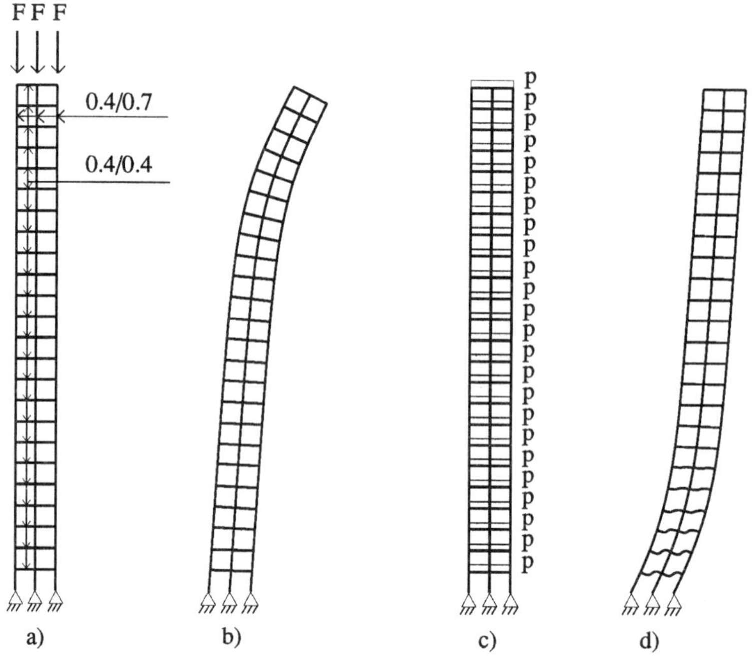

Fig. 5-37. Twenty four storey high framework PF1

With higher frameworks, the situation is similar when the framework is subjected to UDL (Fig. 5-37/d). However, when the load consists of concentrated forces at the top of the columns, the nature of the buckling shape changes. No local shear deformation develops and the framework undergoes combined global bending and shear deformation (Fig. 5-37b). The exact and approximate critical loads and the corresponding error are given in Table 5-12.

Table 5-12. Critical loads for framework PF1

No. of storeys	Critical concentrated load [MN]					Critical UDL [MN]				
	Exact method	Additive method	Δ %	Continuum method	Δ %	Exact method	Additive method	Δ %	Continuum method	Δ %
4	107.10	102.14	5	102.14	5	117.52	104.98	11	127.32	-8
8	105.24	91.34	13	91.34	13	112.67	101.11	10	116.96	-4
12	98.28	77.65	21	77.65	21	110.59	95.26	14	109.13	1
16	79.38	64.19	19	64.19	19	109.25	88.12	19	101.76	7
20	62.28	52.48	16	52.48	16	108.00	80.37	26	94.40	13
24	49.32	42.92	13	42.92	13	105.70	72.57	31	87.09	18

Four-bay framework PF2 has identical columns and beams, respectively (Fig. 5-38). The framework is subjected to five identical concentrated forces at the top of the columns and UDL on the floors.

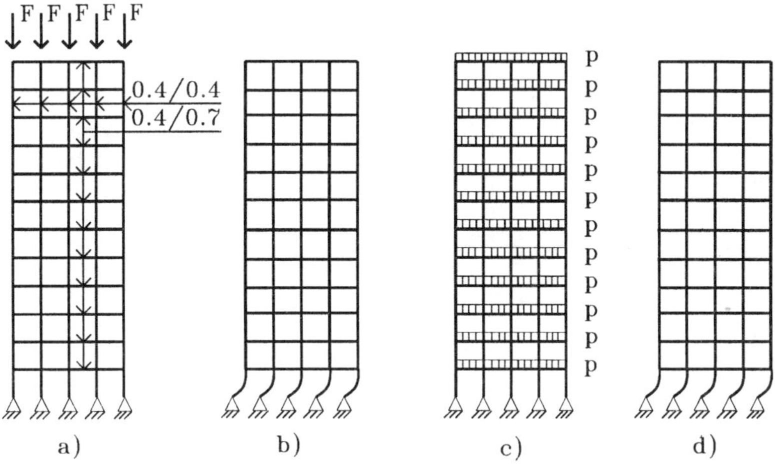

Fig. 5-38. Framework PF2

The stiffness of the beams is greater than that of the columns. The characteristic ratio assumes the value

$$r = \frac{K_c}{K_c + K_b} = \frac{88}{88 + 1829} = 0.046 \ .$$

The buckling shape is determined by local shear deformation at ground floor level in both load cases (Figs 5-38/b and 5-38/d). The critical loads and the error of the approximate procedures are presented in Table 5-13.

Table 5-13. Critical loads for framework PF2

No. of storeys	Critical concentrated load [MN] Exact method	Additive method	Δ %	Continuum method	Δ %	Critical UDL [MN] Exact method	Additive method	Δ %	Continuum method	Δ %
4	83.47	82.78	1	82.78	1	83.30	83.42	-0	84.64	-2
8	83.47	80.09	4	80.09	4	83.30	82.54	1	84.24	-1
12	83.47	75.98	9	75.98	9	83.30	81.12	3	84.03	-1
16	83.47	70.89	15	70.89	15	83.30	79.20	5	83.86	-1
20	83.47	65.26	21	65.26	21	83.30	76.87	8	83.68	-0

Framework PF5 in Fig. 5-39 has very weak columns. The characteristic ratio assumes the value

$$r = \frac{K_c}{K_c + K_b} = \frac{5.48}{5.48 + 341333} = 0.000016 .$$

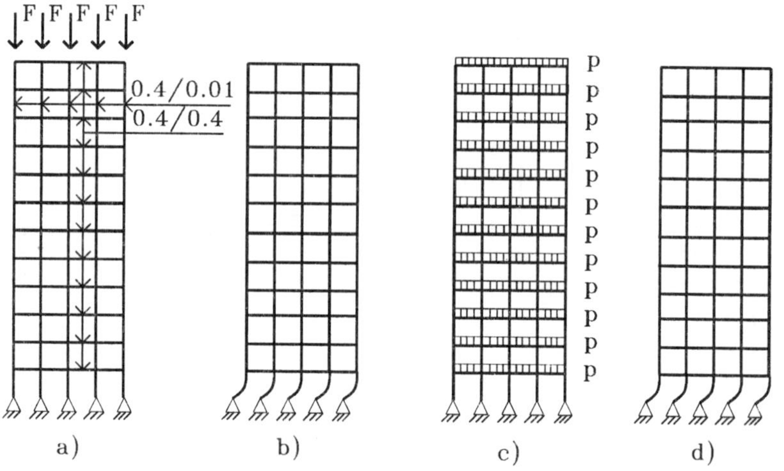

Fig. 5-39. Framework PF5

The buckling shape is again determined by local shear deformation at ground floor level in both load cases (Figs 5-39/b and 5-39/d). The critical loads and the error of the approximate formulae are given in Table 5-14.

Framework PF7 has identical columns and identical beams (Fig. 5-40). The columns are much stronger than the beams. The characteristic stiffness ratio assumes the value

$$r = \frac{K_c}{K_c + K_b} = \frac{87730}{87730 + 5.33} = 0.999939 \ .$$

Table 5-14. Critical loads for framework PF5

No. of storeys	Critical concentrated load [kN]					Critical UDL [kN]				
	Exact method	Additive method	Δ %	Continuum method	Δ %	Exact method	Additive method	Δ %	Continuum method	Δ %
4	1.366	1.371	-0	1.371	-0	1.367	1.371	-0	1.371	-0
8	1.366	1.371	-0	1.371	-0	1.367	1.371	-0	1.371	-0
12	1.366	1.371	-0	1.371	-0	1.367	1.371	-0	1.371	-0
16	1.366	1.371	-0	1.371	-0	1.367	1.371	-0	1.371	-0
20	1.365	1.371	-0	1.371	-0	1.367	1.371	-0	1.371	-0
24	1.365	1.370	-0	1.370	-0	1.366	1.371	-0	1.371	-0

The effect of the weak beams on the deformation of the framework is negligible and the straight columns rotate with respect to the pinned supports in both load cases (Figs 5-40/b and 5-40/d).

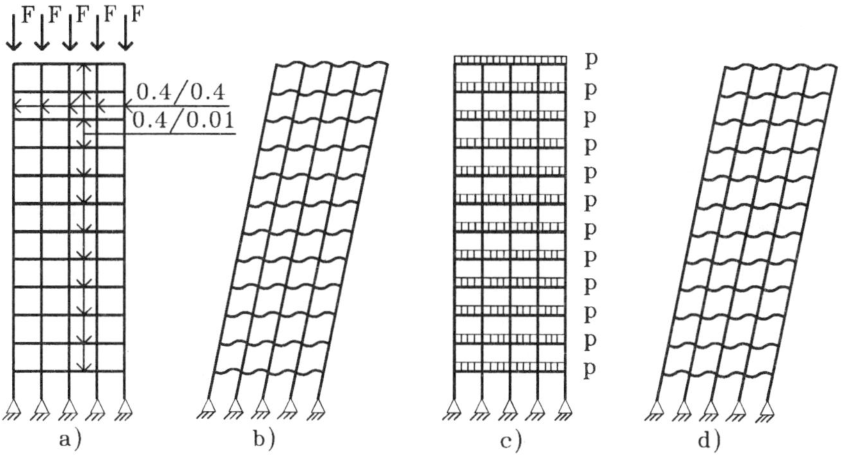

Fig. 5-40. Framework PF7

The critical loads and the corresponding error are presented in Table 5-15.

Table 5-15. Critical loads for framework PF7

No. of storeys	Critical concentrated load [kN]					Critical UDL [kN]				
	Exact method	Additive method	Δ %	Continuum method	Δ %	Exact method	Additive method	Δ %	Continuum method	Δ %
4	5.333	5.333	0	5.333	0	10.666	5.333	50	8.532	20
8	5.333	5.333	0	5.333	0	10.666	5.333	50	9.477	11
12	5.333	5.333	0	5.333	0	10.666	5.333	50	9.838	8
16	5.333	5.333	0	5.333	0	10.666	5.333	50	10.026	6
20	5.333	5.333	0	5.333	0	10.666	5.333	50	10.138	5

As with frameworks on pinned supports, with ground floor beams, the exact critical loads were obtained by relying on elementary equilibrium considerations since the available FE program could not cope with the problem.

An interesting situation emerges with this framework. The last column in Table 5-15 indicates that the the framework under UDL has increasing critical load with increasing height. This is of course not possible. What really happens here is that the continuum method is "over-cautious" and reduces the critical load to a greater extent for lower frameworks than for higher frameworks. As only rigid-body motion develops, the continuum critical load should tend to

$$N_{cr} = 2K = 10.667 \text{ kN}$$

with all the frameworks investigated.

5.6.3 Non-proportional Loading

All the formulae derived in this Chapter assume proportional loading. Proportional loading, however, is not a practical case and certain columns are often overloaded.

Neglecting the effect of overloading, the approximate methods result in overestimated critical loads. As overestimation in stability analysis can easily have fatal effects, it is important to know the extent of the error.

Table 5-16. Critical loads for non-proportional loading

Number of storeys	Continuum method	Exact m. Load "A"	Exact m. Load "B"	Exact m. Load "C"	Exact m. Load "D"	Exact m. Load "E"
4	284.06	285.75	280.80	274.80	274.80	265.50
8	254.24	252.42	248.40	238.80	240.00	238.50
12	216.92	216.00	213.60	201.60	205.20	205.20
16	179.98	180.96	178.80	168.00	172.20	177.45
20	147.67	149.10	147.60	140.40	143.40	147.45

The critical load of framework FF2 of different height, under five different non-proportional load systems, was computed by the continuum method and by the exact method. The load systems and the results are presented in Fig. 5-41 and Table 5-16, respectively.

The results show that the error caused by non-proportionality in loading is not significant: It does not exceed 8 per cent, even with frameworks with significantly overloaded columns.

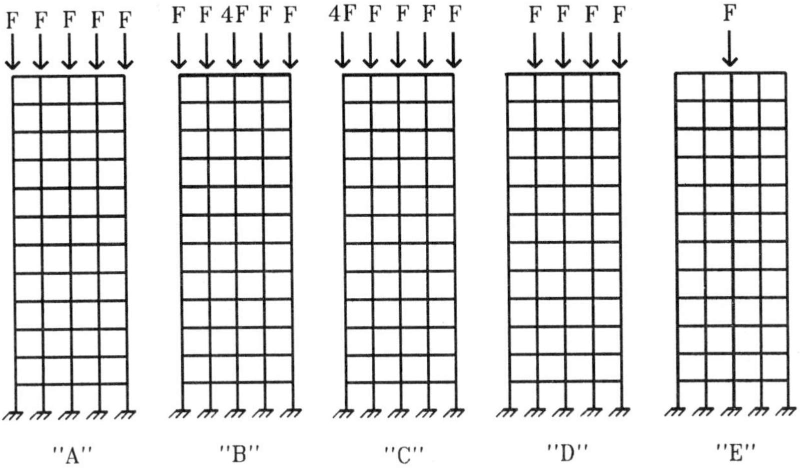

Fig. 5-41. Framework FF2 under non-proportional load systems

The buckling shape of the non-proportionally loaded structures is shown in Fig. 5-42. Due to the load redistribution between the overloaded and the neighbouring (initially) underloaded bars, the buckling shape of the non-proportionally loaded frameworks hardly differs from that of the proportionally loaded cases.

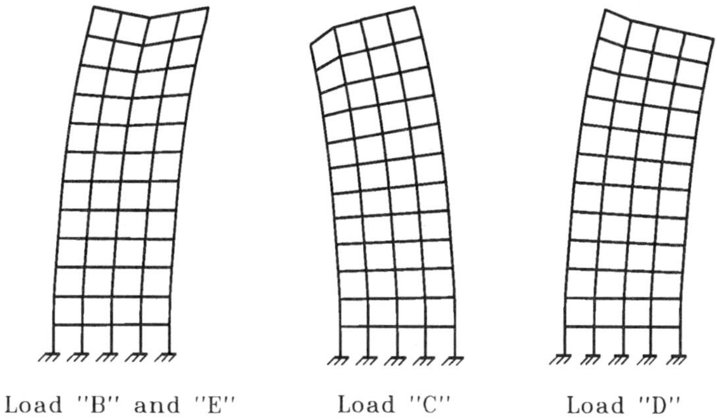

Fig. 5-42. Buckling of non-proportionally loaded frameworks

5.6.4 Test Results

A test was carried out at Budapest Technical University [Szittner, 1979] to check the accuracy of the continuum method of frameworks. The model in Fig. 5-43 is a one-bay, ten-storey non-proportional framework subjected to two concentrated forces on the columns. The critical load of the framework obtained by Southwell's plot [Croll and Walker, 1972], based on the load steps and the corresponding deformations, is

$$F_{cr} = \sum_{1}^{2} F_{cr,i} = 13.90 \text{ kN} .$$

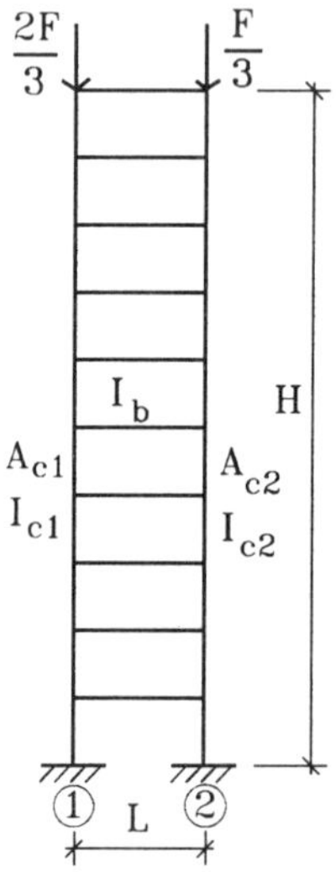

$E = 3200 \text{ N/mm}^2$
$L = 150 \text{ mm}$
$H = 10\times80 = 800 \text{ mm}$
$I_{c1} = 38330 \text{ mm}^2$
$I_{c2} = 10300 \text{ mm}^2$
$A_{c1} = 1878 \text{ mm}^2$
$A_{c2} = 1236 \text{ mm}^2$
$I_b = 3730 \text{ mm}^4$

Fig. 5-43. Geometrical characteristics of small-scale perspex model

The critical load is also calculated from formula (5.24):

$$F_{cr} = \frac{0.95\times0.60 + 11.38 + \dfrac{0.95\times0.60\times11.38}{204.47}}{1 + \dfrac{11.38}{204.47}} = \frac{11.95 + 0.03}{1.0556} = 11.35 \text{ kN.}$$

A computer method based on the exact solution of the problem [Gáspár and Tornyos, 1979] results in

$$F_{cr} = 11.90 \text{ kN}$$

for the critical load. According to the above results, the continuum method offers a conservative estimate. The error made by the application of the continuum method is 4.62 per cent, related to the exact solution and 18.35 per cent, related to the result of the test. A photo of the model is shown in Fig. 5-44.

Fig. 5-44. Testing of the ten-storey perspex model

If the framework is subjected to UDL on the beams, the critical load is obtained in two steps. Neglecting the effect of the excess load at the top, we obtain the critical load from Eqn (5.50):

$$N_{cr} = \frac{0.95\times1.91(12.2-5.97) + 11.38 + \dfrac{0.95\times1.91\times11.38}{649.44}}{1 + \dfrac{0.95\times11.38}{649.44}} =$$

$$= \frac{22.95 + 0.03}{1.0167} = 22.34 \text{ kN} .$$

In making use of Eqs (5.24) and (5.50), Dunkerley's additive formula (5.80) takes the excess load at top into consideration:

$$N_{cr}^{D} = \left(\frac{1}{2\times 10\times 11.35} + \frac{1}{22.34} \right)^{-1} = 20.34 \ .$$

The FE computer procedure, considered exact, yields

$$N_{cr} = 20.88 \text{ kN}.$$

According to these results, the continuum method offers a good and slightly conservative estimate. The error is 2.20 per cent, related to the exact solution.

5.7 CONCLUSIONS

The establishment of the continuum model, the derivation of the differential equations, the solutions at different levels of accuracy and finally the results of the numerical analysis involving hundreds of frameworks of different stiffness ratios all lead to the following conclusions.

1) The continuum method can be applied to the stability analysis of frameworks of four to twenty five storeys, with any number of bays. The procedure is reliable, simple, fast, albeit approximate.

2) As a rule, the method is conservative. The error made by the application of the continuum method is between +20 per cent on the safe side (underestimated critical load) and -10 per cent on the unsafe side (overestimated critical load).

3) The behaviour of frameworks can be characterized by four distinct types of deformation: The local bending deformation of the columns, the global shear deformation, the local shear deformation and the global bending deformation. All four deformations are associated with the corresponding stiffnesses. The overall critical load is determined by a combination of these four stiffnesses.

4) The global shear stiffness of the framework, which is determined by the bending stiffness of the beams, plays a very important role in the response of the structure to the applied vertical load.

5) With frameworks on fixed supports and on pinned supports but with ground floor beams, subjected to concentrated forces on top of the columns, the global bending deformation is dominant, the shear deformation is important and the local bending deformation is negligible (Fig. 5-45).

6) When frameworks on fixed supports and on pinned supports but with ground floor beams are subjected to UDL on the beams, the effect of shear stiffness is dominant (Fig. 5-46).

7) When the frameworks have pinned supports which are not connected with beams, buckling failure develops through the horizontal sway of the ground floor columns in most cases (Fig. 5-47). This, from the point of

view of the framework as a whole, is considered to be local shear deformation. This local shear deformation is closely associated with the critical load of the system of ground floor columns, which determines the critical load of the framework.

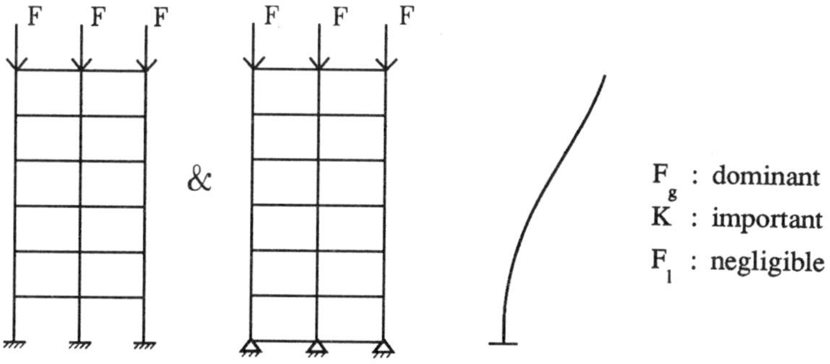

Fig. 5-45. Frameworks subjected to concentrated forces

8) The error made by the application of the classical additive formulae of the theory of stability (Southwell, Föppl-Papkovics) may be considerable: It varies from +50 per cent (overestimated critical load) to -50 per cent (underestimated critical load).

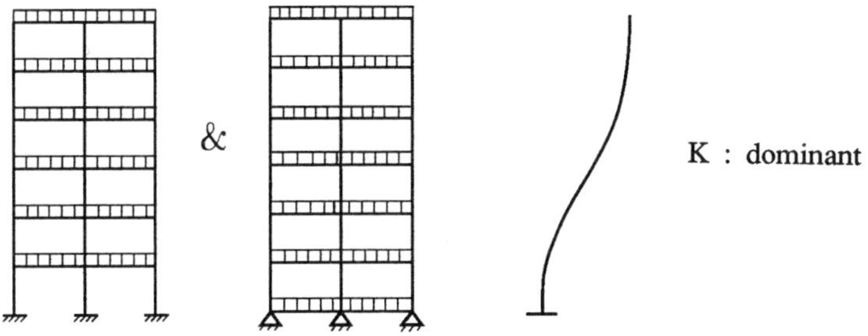

Fig. 5-46. Frameworks under UDL on the beams

9) The continuum method cannot take into consideration the effect of non-proportional loading and results in a slightly overestimated critical load. However, the error is not significant, as a rule. In the cases investigated the error did not exceed 10 per cent, even in cases when individual columns were significantly overloaded.

10) The great advantage of the finite element methods is that they are also applicable to non-regular structures. However, they are sometimes slow and the preparation of the input data may take considerable

time.

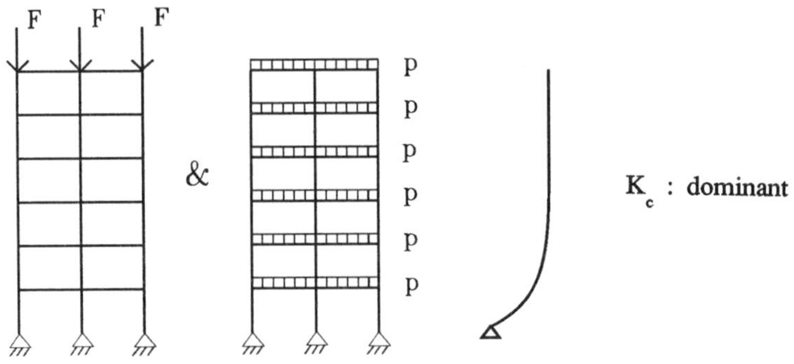

Fig. 5-47. Frameworks on pinned supports

The finite element methods occasionally produce interesting results. In some cases, some FE results indicated greater critical loads for higher frameworks than for lower frameworks, when all the other characteristics of the frameworks were the same. In other cases, errors in the region of 150-250 per cent on the unsafe side (overestimated critical loads) were experienced. However, it should be mentioned that the stiffness characteristics in such cases assumed extreme values. Even so, these cases indicate that great care should be exercised with the FE procedures because the ill-conditioned stiffness matrix might result in faulty critical loads.

6

GENERALIZATION OF THE CONTINUUM METHOD

The continuum method was established for proportional frameworks, however, it is possible to cover structures which are different from the ideal proportional case. These possibilities include frameworks with non-proportional geometrical and stiffness characteristics, coupled shear walls, sandwich columns and cross wall system buildings. It is also possible to investigate structures with semi-rigid beam-column joints. These applications will be presented in this Chapter.

Although generalization is possible in many cases, it is important to establish the limits of possible application [Hegedüs, 1987].

6.1 NON-PROPORTIONAL FRAMEWORKS

All the formulae derived so far are applicable to proportional structures. However, the formulae can easily be extended to cover non-proportional and even non-regular cases.

6.1.1 Frameworks with Columns of Different Cross-section

A structure is non-proportional, if all the columns have the same cross-section and non-regular if the cross-sections are different. Formula (5.8) automatically takes these cases into consideration.

If the length and/or the cross-section of the beams is different in the bays - and this leads to non-regularity in most cases -, then formula (5.5) should be replaced by the more general formula

$$k_{b,i} = \frac{6E_b I_{b,i}}{l_i h} , \tag{6.1}$$

which stands for the stiffness of the i-th beam. Formula (5.9) in Section 5.3.1.1 then automatically produces the sum of the stiffness of the beams.

6.1.2 Frameworks on Pinned Supports without Ground Floor Beams

It is a common case that the ground floor level beams connecting the pinned supports are missing (Fig. 6-1/a) or are not strong enough to be taken into account. This case is only different from the one investigated in detail in Section 5.4 in that the first storey region is the most vulnerable from the point of view of additional local sway (Fig. 6-1/b). In assuming beams with infinitely great stiffness, failure is reached through the local sway of the first storey region (Fig. 6-1/c). The effective length of the first storey columns is

$$h^{*} = 2h \tag{6.2}$$

and the formula of the critical load that belongs to the local sway, i. e., the local shear stiffness assumes the form

$$K_{c} = \sum_{1}^{n} k_{c,i} = \frac{\pi^{2} E_{c} I_{c}}{h^{*2}} , \tag{6.3}$$

where I_c is the sum of the second moment of area of the columns.

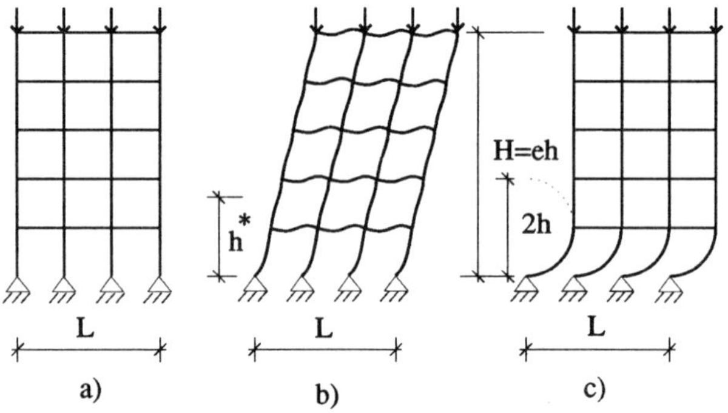

Fig. 6-1. Frameworks on pinned supports, without ground floor beams: a) non-proportional frame, b) practical case, c) stiff beams

All the other formulae derived in Section 5.4 for frameworks on pinned supports can be used except for (5.53) which should be replaced by formula (6.3). The formula for the critical load for frameworks subjected to concentrated forces assumes the form

$$F_{cr} = \frac{K}{1 + \dfrac{K}{F_{g}}} . \tag{6.4}$$

When the framework is subjected to uniformly distributed load at floor levels, the continuum formula assumes the form

$$N_{cr} = \frac{(\alpha-\beta)rN_1^* + K}{1 + \dfrac{rK}{N_g}}. \qquad (6.5)$$

It should be borne in mind that, because of the missing ground floor beams, the local shear stiffness is to be computed from formula (6.3). All the other terms in formula (6.5) are given and explained in Section 5.4.

As with frameworks on fixed supports, subjected to UDL at floor levels, the continuum formula for the critical load may result in a critical load greater than the exact one. The reason for this discrepancy is two-fold. First, as the vertical load increases from the top of the framework to ground floor level, the stiffness of the columns is constant. Second, when the substitute column is created and the load consisting of concentrated forces at floor levels is distributed uniformly along the column, a "surplus" concentrated force at the top of the substitute column is not allowed for. (See also explanation in the introduction of Section 5.3.2.) By applying Dunkerley's reciprocal formula to the critical load given by formula (6.5) and to the "surplus" concentrated load, the latter approximation can be allowed for in the same way as in Section 5.5.2. The critical load can thus be computed from

$$N_{cr}^D = \left(\frac{r}{2eF_{cr}} + \frac{1}{N_{cr}}\right)^{-1}, \qquad (6.6)$$

where the critical loads are given by expressions (6.4) and (6.5). Factor e in the formula stands for the number of storeys. The introduction of reduction factor r defined by formula (5.23) into formula (6.6) controls the effect of Dunkerley's correction. When the value of r decreases, the effect of Dunkerley's correction also decreases. As the numerical examples in Section 5.6.2 show, the buckling shape of the columns of the framework with small r tends to a straight vertical line above the first storey region. The distribution and the location of the applied load is no longer important in such cases and indeed reduction factor r in formula (6.6) reduces - or even eliminates - the effect of Dunkerley's correction.

6.1.3 Frameworks with Longer First Storey Columns

Again, it is a practical case that the columns at ground floor level are longer than the ones at the other storey levels (Fig. 6-2/a/b/c). The behaviour of such frameworks only differs from the corresponding proportional ones in local shear buckling. Because of the "softer" first storey region, local failure would occur through the buckling of the ground floor columns. Formula (6.3) can also be used with these frameworks but with

$$h^* = 2\bar{h} \tag{6.7}$$

for the framework in Fig. 6-2/a and with

$$h^* = \bar{h} \tag{6.8}$$

for the frameworks in Figs 6-2/b and 6-2/c. The critical load is then obtained by applying the corresponding formulae of (6.4), (6.5) and (6.6). Parameter $\bar{h}$ in the above formulae marks the length of the longer first storey columns.

Because of the longer first storey columns, however, the approximation made when the UDL is distributed along the substitute column has a greater - unfavourable - effect on the critical load. The difference between the centre of the applied load (UDL at floor levels) and the centre of the substitute load of the substitute column (UDL along the column) is greater as with the equivalent proportional frameworks, especially with the frameworks shown in Figs 6-2/a and 6-2/c.

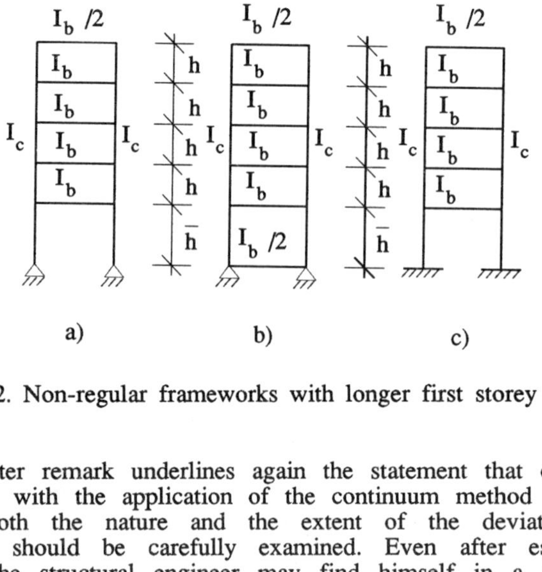

Fig. 6-2. Non-regular frameworks with longer first storey columns

This latter remark underlines again the statement that one should be very cautious with the application of the continuum method to non-regular structures. Both the nature and the extent of the deviation from the regular case should be carefully examined. Even after establishing the differences, the structural engineer may find himself in a situation when neither the magnitude nor the sign of the error is known. The results of numerical examples to show the accuracy and the limitation of the continuum method are presented in Section 5.6.

6.2 COUPLED SHEAR WALLS

Shear walls and coupled shear wall systems represent another large group of structures used in the building industry. Above a certain height, frameworks alone cannot ensure the necessary lateral stiffness of a building and a system of coupled shear walls is often introduced.

The exact analysis of coupled shear walls is a complex problem. The exact solution is only available for special cases and therefore is mainly of theoretical interest. The approximate methods applied for the analysis of coupled shear walls can be divided into two major groups, depending on the model they are based on. Methods using finite elements make it possible to investigate structures of practically arbitrary shapes and boundary conditions. Even subtleties such as change in material or in wall thickness, initial stresses etc. can be allowed for. However, this versatility can soon put the structural engineer beyond the capacity of the available computer, not to mention the time consuming process of inputting data. Analytical methods, on the other hand, offer simple solutions for regular structures in many cases.

Well developed and publicized methods are available for the *stress* analysis of coupled shear walls subjected to horizontal load [Rosman, 1965a; Rosman, 1965b; MacLeod, 1967; Stafford-Smith and Coull, 1972; Schueller, 1977; Irwin, 1984; Taranath, 1988]. The *stability* analysis of large coupled shear walls, however is a less developed area and the available simple procedures have considerable limitations [Beck and König, 1966; Beck, 1972; König, Ambos, Bubenheim and Heunish, 1972]. Rosman [1973] developed a simple procedure, making use of stability parameters, arranged in tables. His method, however, is only applicable to single-bay structures.

As is shown later on, coupled shear walls can be considered frameworks with special stiffness and geometrical characteristics so that the continuum method originally developed to frameworks can also be applied to coupled shear walls.

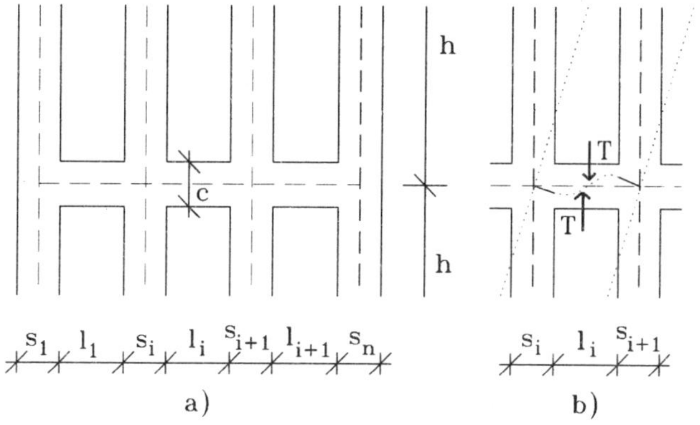

Fig. 6-3. Coupled shear walls: a) one storey section, b) shear force T

6.2.1 Coupled Shear Walls Subjected to Concentrated Forces

Relying on the continuum procedure developed for frameworks in Chapter 5, the stability analysis of coupled shear walls can be carried out easily as formulae of closed form can be derived for the critical load. A system of coupled shear walls can be regarded as a special framework where walls (instead of columns) are connected with beams or lintels (Fig. 6-3/a). However, there are two basic differences between coupled shear walls and frameworks.

a) The two end-sections of the beams cannot develop bending because the walls, whose stiffness is considered infinitely great along these sections, do not let them. The centroidal axis of these sections must be characterized by straight lines when the deflection curve of the beams is established. These straight sections are marked with solid lines in Fig. 6-3/b. It is interesting to note that this is also the case with frameworks but because of the relatively short sections in the case of frameworks, this fact is not taken into account in the analysis of frameworks in structural engineering practice.

b) As a rule, in the case of coupled shear walls, the depth of the cross-section of the beams is relatively great and the beams are relatively short. In other words, the beams of coupled shear walls are not as slender as those of the frameworks and therefore the shear deformation of the beams is no longer negligible.

Bearing in mind the above differences, the equations and formulae derived for the stability analysis of frameworks can be used for the stability analysis of coupled shear walls if certain modifications are made. Considerations must be given to the increased stiffness at the nodes as well as to shear deformations resulting from the small span-to-depth ratios of the lintels. Both differences only affect the shear stiffness and the necessary modifications can be built into formula (5.9).

The distributed bending stiffness of the beams given by Eqn (5.5) in Section 5.3.1 was derived on the assumption that the translation of the point of contraflexure of the beam is zero (Fig. 5-4). As the ratio of depth per length of the beams of coupled shear walls is often much greater than the same ratio with the beams of frameworks, the translation due to shear deformation can be of considerable magnitude and should be built into the equation expressing the translation of the point of contraflexure:

$$-y'\frac{l_i}{2} + \frac{T(l_i/2)^3}{3E_bI_b} + \frac{T(l_i/2)}{GA_b/\rho} = 0\ . \tag{6.9}$$

In Eqn (6.9) the following notation is used:

G	modulus of elasticity in shear of the beams,
A_b	cross-sectional area of the beams,
l_i	distance between the i-th and (i+1)-th walls (Fig. 6-3),
ρ	constant depending on the shape of the cross-section (ρ=1.2 for rectangular cross-sections).

After some mathematical manipulation, formulae (5.1) and (6.9) can be converted into the same form [Zalka, 1982a] by modifying the second moment of area of the cross-section of the beams:

$$I_b^* = \frac{I_b}{1 + 12\dfrac{\rho E_b I_b}{l_i^2 GA_b}} \; . \tag{6.10}$$

In formula (6.10) I_b is the "original" second moment of area of the cross-section and I_b^* is the modified second moment of area which also takes the effect of the shear deformation into account.

The deflection curve of the beam of a shear wall system defers from the deflection curve of a beam of a framework. The difference is due to the fact that there are straight sections in the curve where the beam joins the wall (Fig. 6-3/b). All one has to do is re-formulate condition (5.1) in Section 5.3.1, expressing that the translation of the contra-flexure point of the beam is zero. The new equation related to the *i*-th beam assumes the form

$$-y'\left(\frac{l_i}{2} + \frac{s_i}{2}\right) + \frac{T(l_i/2)^3}{3E_b I_b^*} = 0 \; . \tag{6.11}$$

In this equation s_i is the width of the *i*-th wall and I_b^* is the second moment of area of the beam defined by (6.10), which takes the shear deformation of the beam into account. After some rearrangement of Eqn (6.11) and making use of Eqs (5.2), (5.3) and (5.4), we obtain

$$k_{b,i}^l = \frac{6E_b I_b^*(l_i + s_i)^2}{l_i^3 h} \tag{6.12}$$

for the left-end beam stiffness. The similar formula

$$k_{b,i}^r = \frac{6E_b I_b^*(l_i + s_{i+1})^2}{l_i^3 h} \tag{6.13}$$

stands for the right-end beam stiffness where s_{i+1} is the width of the (i+1)-th wall.

As for the additional local and global sway of the columns, all that was said earlier about frameworks holds. It follows that formula (5.24) can still be used but with a shear stiffness modified according to the foregoing:

$$F_{cr} = \frac{rF_1 + K^* + \dfrac{rF_1 K^*}{F_g}}{1 + \dfrac{K^*}{F_g}} = rF_1 + \left(\frac{1}{F_g} + \frac{1}{K^*} \right)^{-1}. \tag{6.14}$$

The modified shear stiffness in formula (6.14) is defined by

$$K^* = K_b^* \frac{K_c}{K_b^* + K_c} \tag{6.15}$$

where

$$K_b^* = \sum_1^{n-1} \frac{6E_b I_{b,i} \left((l_i + s_i)^2 + (l_i + s_{i+1})^2 \right)}{l_i^3 h \left(1 + 12 \dfrac{\rho E_b I_b}{l_i^2 G A_b} \right)} \tag{6.16}$$

is the distributed stiffness of the beams. All the other parameters needed for formula (6.14) are given and explained in detail in Section 5.3.1.

The repeated application of Southwell's additive formula and the Föppl-Papkovics reciprocal additive formula results in

$$F_{cr} = \frac{F_1 + K^* + \dfrac{F_1 K^*}{F_g}}{1 + \dfrac{K^*}{K_g}} = F_1 + \left(\frac{1}{F_g} + \frac{1}{K^*} \right)^{-1} \tag{6.17}$$

for the critical load.

6.2.2 Coupled Shear Walls with UDL at Floor Levels

When coupled shear walls subjected to concentrated forces were investigated in Section 6.2.1, it was shown that the formulae derived for frameworks could also be applied to coupled shear walls if certain modifications were made. These modifications do not depend on the nature of the loading and only affect the shear stiffness of the structure. With coupled shear walls subjected to UDL at floor levels, the formula of the modified shear stiffness (Eqn 6.16) should be combined with the formula derived in Chapter 5 for frameworks subjected to UDL (Eqn 5.50). By so doing, the formula for the critical load of coupled shear walls subjected to uniformly distributed load at floor levels assumes the form

$$N_{cr} = \frac{(\alpha-\beta)rN_1 + K^* + \dfrac{rN_1K^*}{N_g}}{1 + \dfrac{K^*}{N_g}} \tag{6.18}$$

where K^* is defined by expressions (6.15) and (6.16). The other factors in formula (6.17) are defined in Section 5.3.2, including the critical load parameter α, which is given in Table 5-2 and Figs (5-15) and (5-16), in the function of the stiffness ratio

$$\beta = \frac{K^*}{N_1} . \tag{6.19}$$

The repeated application of Southwell's additive formula and the Föppl-Papkovics reciprocal additive formula now leads to formula

$$N_{cr} = \frac{N_1 + K^* + \dfrac{N_1K^*}{N_g}}{1 + \dfrac{K^*}{N_g}} = N_1 + \left(\frac{1}{N_g} + \frac{1}{K^*}\right)^{-1} \tag{6.20}$$

for the critical load.

As with frameworks on fixed supports, under UDL on the beams, the accuracy of the continuum formula of the critical load can be increased by taking the effect of the "excess" concentrated load at the top of the structure into consideration. This can be done by the application of Dunkerley's formula. By so doing, the formula

$$N^D_{cr} = \left(\frac{1}{2eF_{cr}} + \frac{1}{N_{cr}}\right)^{-1}, \tag{6.21}$$

is obtained, where the continuum critical loads are defined by Eqs (6.14) and (6.18) and e stands for the number of storeys.

It should be pointed out here that parameter l with coupled shear walls defines the length of the beams (the distance between the walls) - and not the distance between the wall axes.

6.2.3 Design Examples - Accuracy Analysis

To check the accuracy of the formulae, a numerical analysis is carried out, involving three sets of coupled shear wall systems, as design examples. The structures for the numerical analysis are subjected to both concentrated forces at the top of the walls and UDL at floor levels. The size of the coupled shear wall systems varies from four to twenty four storeys, with different stiffness characteristics. The values of the modulus of elasticity and the shear modulus are $E=3\text{x}10^4$ MN/m^2 and $G=1.25\text{x}10^4$ MN/m^2, respectively. The thickness of the wall systems is 0.2 m in all three cases. The storey height is 3 m. The nature of the buckling shape of the coupled shear wall systems is also given in the figures.

The critical loads computed by a sophisticated FE procedure, considered exact, by the continuum method and by the repeated application of Southwell's theorem and the Föppl-Papkovics theorem are presented in tables. The tables also give the error in per cents, made by the application of the approximate formulae. Positive error means conservative estimate and the error is marked negative when the approximate critical load is greater than the exact one.

The results of a small scale test are also presented.

6.2.3.1 Comparison with FE Results

Coupled shear wall system CSW1 in Fig. 6-4/a represent a characteristic arrangement when the shear wall is pierced by two rows of window openings.

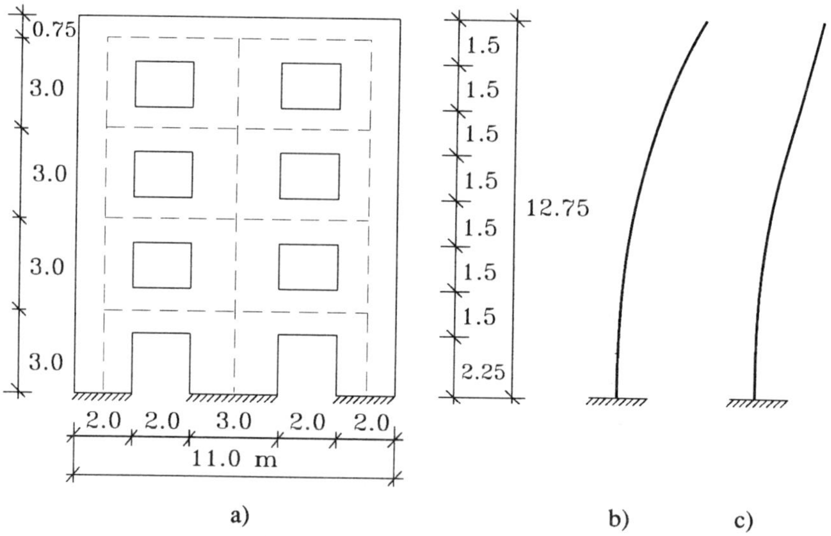

Fig. 6-4. a) Shear wall CSW1 with two bands of window openings and buckling shape with b) concentrated load, c) UDL at storey levels

The stiffness of the wall in the middle is greater than that of the side ones. The stiffness of the connecting lintels is considerably great. The system of walls and lintels with relatively great cross-section does not even resemble a framework. Still, the continuum formulae, based on the analysis of frameworks, derived for the critical load of coupled shear walls result in good approximations. The buckling shape is dominated by bending in both loading cases (Figs 6-4/b and 6-4/c), with some contribution from global shear deformation when the structure is under UDL (Fig. 6-4/c). The exact and approximate critical loads and the error made by the application of the approximate formulae are given in Table 6-1.

Table 6-1. Critical loads for coupled shear wall system CSW1

No. of storeys	Critical concentrated load [MN]					Critical UDL [MN]				
	Exact method	Additive method	Δ %	Continuum method	Δ %	Exact method	Additive method	Δ %	Continuum method	Δ %
4	5754.0	4565.9	21	4433.7	23	11424.0	7582.1	34	8097.8	29
8	1923.0	1763.0	8	1729.9	10	4876.8	4004.8	18	4245.6	13
12	912.0	875.0	4	860.3	6	2505.6	2311.4	8	2423.1	3
16	525.0	513.3	2	505.1	4	1505.3	1456.1	3	1518.0	-1
20	339.0	335.2	1	329.9	3	1002.0	987.2	2	1025.6	-2
24	237.0	235.4	1	231.7	3	706.3	708.5	-0	734.0	-4

Coupled shear wall system CSW2 in Fig. 6-5/a has two rows of door openings. The stiffness of the beams connecting the walls are now by orders of magnitude smaller than that of the walls. The length of the two sets of beams is different.

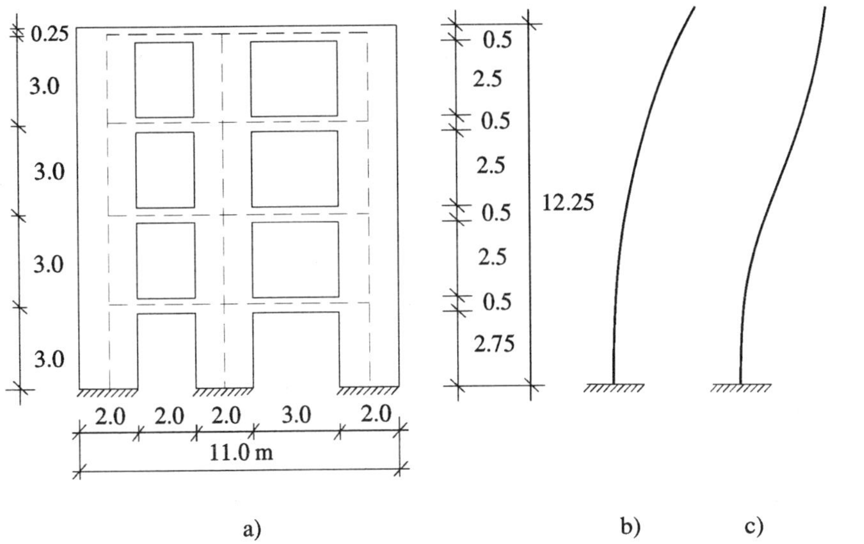

Fig. 6-5. a) Shear wall CSW2 with two bands of door openings and buckling mode with b) concentrated load at top, c) UDL at storey levels

Table 6-2. Critical loads for coupled shear wall system CSW2

No. of storeys	Critical concentrated load [MN]					Critical UDL [MN]				
	Exact method	Additive method	Δ %	Continuum method	Δ %	Exact method	Additive method	Δ %	Continuum method	Δ %
4	918.0	772.9	16	763.4	17	1838.4	1248.0	32	1576.5	14
8	553.0	522.7	6	520.4	6	1146.2	720.7	37	1104.3	4
12	390.0	390.6	-0	389.5	0	858.2	577.1	33	855.2	0
16	289.7	294.0	-2	293.4	-1	680.2	486.1	29	687.8	-1
20	219.9	224.1	-2	223.7	-2	546.9	413.0	25	562.2	-3
24	170.5	173.8	-2	173.6	-2	444.6	351.6	21	464.2	-4

The buckling shape is characterized by the bending of the walls in both loading cases, with considerable contribution from global shear deformation when the structure is under UDL (Figs 6-5/b and 6-5/c). The exact and approximate critical loads and the corresponding error are given in Table 6-2.

Fig. 6-6 shows coupled shear wall system CSW3, consisting of four shear walls connected by beams at floor levels. This structure is nearly proportional. The second moment of area of the walls is the same along the height but the second moment of area of the side walls is approximately half of that of the inner ones. All the beams have the same cross-section except the upmost ones whose second moment of area is approximately half of that of the others.

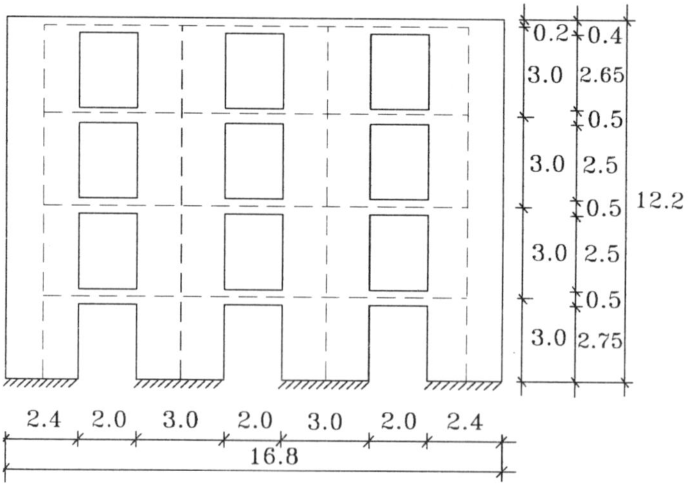

Fig. 6-6. Coupled shear wall system CSW3

The exact and approximate critical loads and the corresponding error are given in Table 6-3. The buckling shape of the system is characterised by the bending of the walls and by global shear deformation in both load cases, exactly in the same way as with CSW2 (Figs 6-5/b and 6-5/c).

Table 6-3. Critical loads for coupled shear wall system CSW3

No. of storeys	Critical concentrated load [MN]					Critical UDL [MN]				
	Exact method	Additive method	Δ %	Continuum method	Δ %	Exact method	Additive method	Δ %	Continuum method	Δ %
4	2971.4	2364.2	20	2336.6	21	5889.6	3953.6	33	4881.5	17
8	1876.0	1598.1	15	1591.2	15	3808.0	2196.1	42	3409.6	11
12	1372.1	1223.9	11	1220.8	11	2908.8	1755.2	40	2659.8	9
16	1029.4	944.4	8	942.7	8	2345.6	1494.1	36	2163.7	8
20	783.7	734.2	6	733.1	7	1911.2	1287.7	33	1792.4	6
24	607.7	578.2	5	577.4	5	1564.8	1112.4	29	1499.3	4

6.2.3.2 *Test Results*

A small-scale perspex model was tested at Budapest Technical University [Szittner, 1979] to give information of the accuracy of the formula derived for the stability analysis of coupled shear walls. A one-bay, eleven-storey, proportional system of shear walls subjected to two concentrated forces was investigated (Fig. 6-7).

Southwell's plot [Croll and Walker, 1972] in this case gives

$$F_{cr} = \sum_1^2 F_{cr,i} = 10.75 \text{ kN}$$

for the critical load. The value

$$F_{cr} = \frac{0.99\times3.36 + 12.44 + \dfrac{0.99\times3.36\times12.44}{36.75}}{1 + \dfrac{12.44}{36.75}} = \frac{15.78 + 1.13}{1.3385} = 12.63 \text{ kN}$$

is obtained from formula (6.14) derived for coupled shear walls. The computer program based on the exact solution yields

$$F_{cr} = 13.60 \text{ kN}$$

for the critical load. The deviation of the continuum solution from the exact one is 7.13.

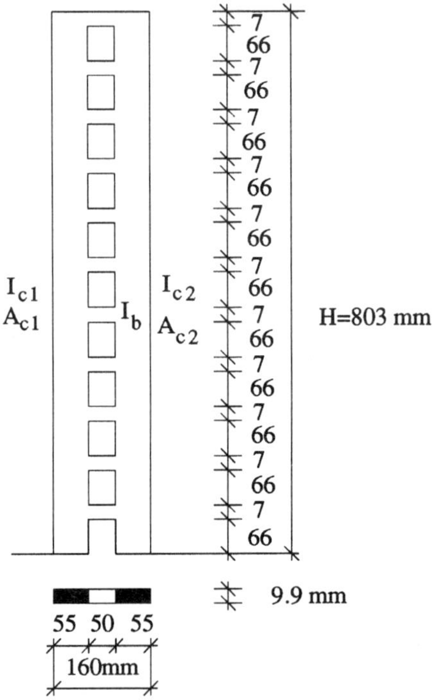

Fig. 6-7. Coupled shear walls model

If the coupled shear wall shown in Fig. 6-7 is subjected to UDL at floor levels, the critical load is obtained from formula (6.18) and (6.21):

$$N_{cr} = 34.13 \text{ kN} .$$

Rosman's [1973] method, which is only applicable to one-bay, symmetric coupled shear walls offers useful comparison. According to his method, the critical load is

$$N_{cr} = 34.37 \text{ kN} .$$

The exact solution is

$$N_{cr} = 33.53 \text{ kN} .$$

The critical load obtained by the continuum method is now 1.79 per cent greater than the exact value. The model is shown in Fig. 6-8.

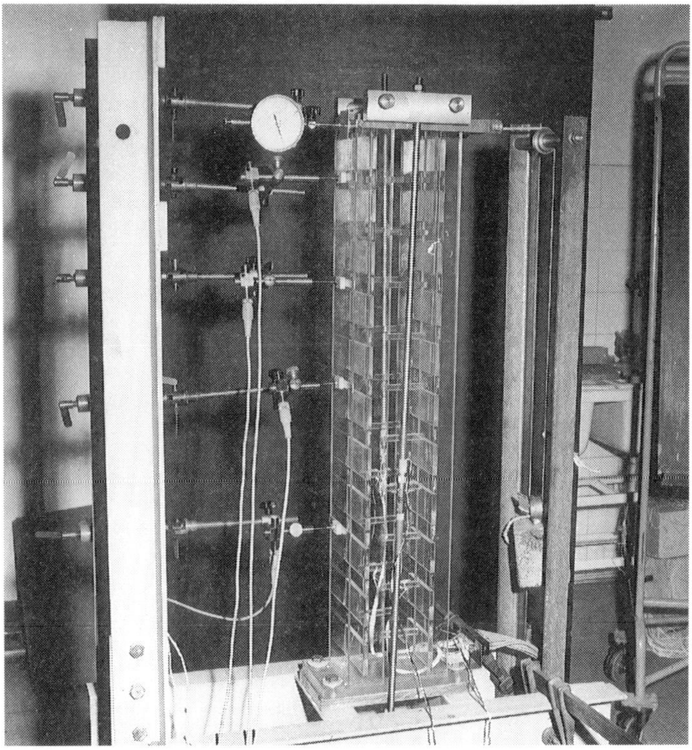

Fig. 6-8. Perspex model of coupled shear walls

6.2.4 Conclusions

The results of the accuracy analysis show that the continuum method offers reliable approximate values for the critical load. As with frameworks, the analysis can be carried out very easily and in minutes.

The method is approximate. In the cases investigated the error was between +30 per cent (underestimated critical loads) and -10 per cent (overestimated critical loads). For structures between 8 to 20 storeys, the most conservative estimates dropped bellow +15 per cent.

Of the four characteristic deformations, global bending and global shear deformations - and the corresponding critical loads - play an important role. Local bending deformation - and the corresponding critical load - only have a minor effect on the overall critical load. Local shear deformation, i.e., sway between the floor levels is negligible, as a rule.

It is worth mentioning that some well-publicized FE procedures have difficulties with the stability analysis in certain stiffness regions when the system is too large. Incorrect load factors were obtained for different twelve storey coupled shear wall systems with two rows of openings. As the stiffness matrix is often ill-conditioned to a different extent with coupled shear walls, it is sometimes difficult even to realize that the results are not reliable. With the twelve storey structures investigated, the incorrect buckling shape was a warning sign. However, when the buckling shape looks reasonable, it is very difficult indeed to find out that the critical load may not be correct.

According to detailed investigations into the solution of the eigenvalue problem characterizing the stability of coupled shear wall systems, the reason for the incorrect critical load factors is that the first eigenvalue is sometimes overlooked by the step-by-step procedures [Zalka, 1982b].

6.3 CROSS WALL SYSTEM BUILDINGS

When the centre of the support system of a building and that of the applied vertical load coincide, stability failure may occur in three different ways. The system can develop in-plane buckling in both principal planes and pure torsional buckling. The formulae presented in Chapter 5 and Section 6.2 make it possible to check the stability of cross wall system buildings, which consist of parallel frameworks and coupled shear walls in a symmetric arrangement. Planar stability analysis can be carried out, when the elements of the system develop buckling in their plane. The floors are assumed to be infinitely stiff in their plane and infinitely flexible perpendicular to their plane. Figure 6-9 shows the layout of such a cross wall system building, where

S_1 and S_2	are the size of the ground plan [m],
s	is the distance between the support elements [m],
m	is the number of support elements in the building.

6.3.1 Approximate Solution

According to Southwell's additive theorem, the sum of the critical loads of the support elements (frameworks and coupled shear walls) is a lower bound to the critical load of the system. Formula

$$F_{cr}^{bu} = \sum_{1}^{m} F_{cr,j} \tag{6.22}$$

and

$$N_{cr}^{bu} = \sum_{1}^{m} N_{cr,j} \tag{6.23}$$

give the total critical load of cross wall buildings subjected to concentrated forces at top floor level and uniformly distributed load at floor levels. Parameter m in (6.22) and (6.23) is the number of the frameworks and coupled shear walls supporting the building and $F_{cr,j}$ and $N_{cr,j}$ are the critical loads of the individual elements.

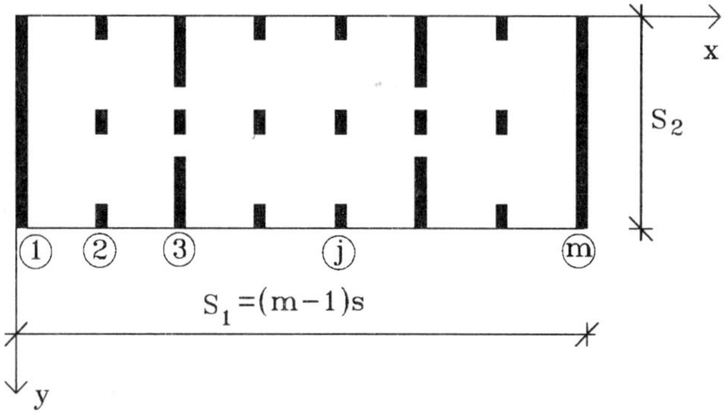

Fig. 6-9. Layout of a cross wall system building

The fact that the frameworks and the coupled shear walls develop different type of deformation - shear and bending type deformation (Fig. 6-10) is not favourable, as far as the accuracy of the formulae is concerned. This difference in deformation often has a considerable effect on the accuracy of the *stress* analysis. With the stability analysis, however, the difference is not significant and formulae (6.22) and (6.23) are approximate anyway.

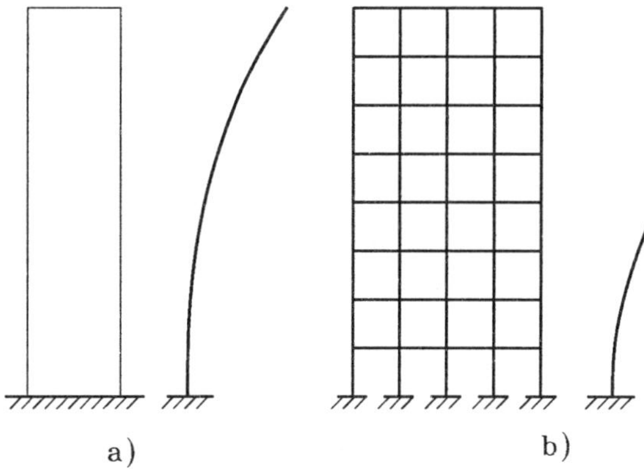

Fig. 6-10. Characteristic deformation of a) shear wall and b) frame

By applying Dunkerley's reciprocal formula to buildings subjected to both concentrated forces at top floor level and UDL at floor level, the ratio of the overall critical load and the applied loads can be produced:

$$\nu = \frac{1}{\dfrac{\sum_{1}^{m} F_j}{F_{cr}^{bu}} + \dfrac{QS_1S_2e}{N_{cr}^{bu}}} . \qquad (6.24)$$

In formula (6.24),

Q is the intensity of the UDL on the floors [kN/m^2],

e is the number of storeys,

and

$$\sum_{1}^{m} F_j$$

is the sum of the concentrated forces acting at the top of the frameworks and coupled shear walls.

This approximate formula offers useful information about the safety of a cross wall system building: It shows that, with the fixed ratio of the applied concentrated load and the UDL, what the ratio of the critical load of the building to the total applied load is, as far as planar buckling parallel to the cross walls and frameworks is concerned. A computer procedure based on the method introduced here is available [Fazakas et al, 1988]. The procedure makes it possible for the structural engineer to carry out rapid and comprehensive checks in minutes, producing an economic structural solution.

The remarks made in Section 5.5.3 on overloading certain columns of a framework subjected to concentrated forces still hold but this time overloading also refer to the overloading of the individual frameworks and coupled shear walls. The factor of maximum overloading must not exceed four.

The building shown in Fig. 6-9 may not be stable in direction x. Stability in this direction should also be ensured. This can be done by perpendicular (coupled) shear walls or by cores. In the first case a stability analysis, similar to the one just presented, can be carried out. Buildings supported by cores and developing torsional buckling are investigated in Chapter 7.

6.3.2 Design Example

To show how the formulae can be applied to structural engineering design, compute the ratio of the critical load to the applied load of the twenty storey cross wall system building in Fig. 6-11/a. The building is supported by five frameworks and two systems of coupled shear walls in symmetric arrangement. Shear wall system CSW1 is known from Section 6.2.3.1. The geometrical and stiffness characteristics are given in Fig. 6-4.

Framework FF9 is also twenty storeys high, with a storey height of

3 m (Fig. 6-11/b). The cross-section of all the beams and columns is $0.4 \times 0.4 = 0.16$ m^2. The length of the beams is 5.3 m. The shear modulus is 12,500 MN/m^2 and the modulus of elasticity is 30,000 MN/m^2. The columns are 3.0 m long.

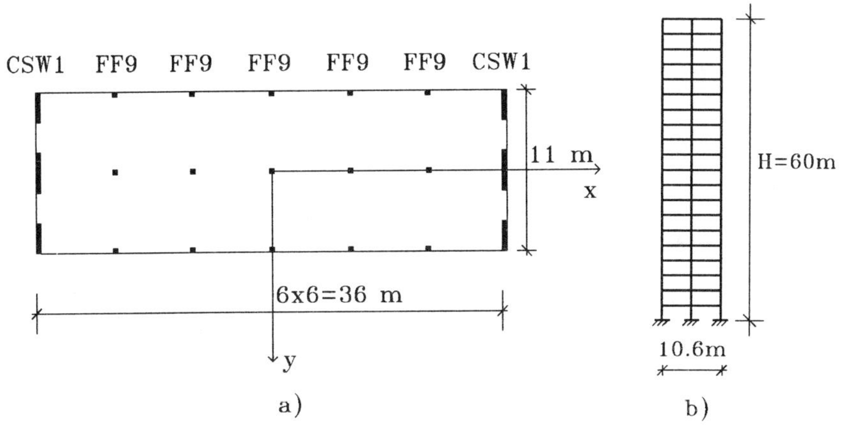

Fig. 6-11. Cross wall system building for the design example a) layout and b) framework FF9

The building is subjected to UDL on the floors:

$$Q = 15 \text{ KN/m}^2$$

and concentrated forces at the top of the building. The sum of the concentrated forces is

$$\sum_1^m F_j = 10 \text{ MN}.$$

First, the critical loads of the support elements are needed:

CSW1: $F_{cr,1} = 329.93$ MN, $N_{cr,1} = 1025.55$ MN,

FF9: $F_{cr,2} = 48.84$ MN, $N_{cr,2} = 70.37$ MN,

The above critical loads are computed from formulae (6.14), (6.20), (6.21), (5.24), (5.50) and (5.80), respectively. The critical load of the building for the concentrated load is then obtained from formula (6.22):

$$F_{cr}^{bu} = \sum^m F_{cr,j} = 904.1 \text{ MN}.$$

When the building is under UDL on the floors, the critical load is obtained from formula (6.23):

$$N_{cr}^{bu} = \sum_{1}^{m} N_{cr,j} = 2402.9 \text{ MN} .$$

Formula (6.24) takes both loads into consideration and gives the ratio of the overall critical load to the applied loads:

$$\nu = 16.5 .$$

According to this result, the building can theoretically withstand 16.5 times as much load as the currently applied load. For the actual structural design, the application of safety factors in accordance with the relevant code of practice may have to be considered.

To check the accuracy of the above results, the critical loads are also computed by a sophisticated finite element procedure. The FE procedure also assumes infinitely stiff floor slabs, otherwise it is considered to be exact. After a lengthy preparation and input of data, the critical loads emerge as

$$F_{cr}^{bu} = \sum_{1}^{m} F_{cr,j} = 910.2 \text{ MN}$$

for the concentrated load and

$$N_{cr}^{bu} = \sum_{1}^{m} N_{cr,j} = 2362.8 \text{ MN} .$$

for the UDL. The ratio of the overall critical load to the applied loads turns out to be

$$\nu = 16.3 .$$

6.4 ANALOGY BETWEEN FRAMEWORKS AND SANDWICH COLUMNS

Certain similarities indicate that for uniformly distributed normal loads, there is an analogy between the stability analysis of large frameworks and that of sandwich columns (cantilevers developing both bending and shear deformations). This analogy can clearly be seen if the approximate procedures used in solving the two problems are compared. First, the governing differential equation of the cantilever developing both bending and shear deformation will be established. Second, the approximate solution will be presented and compared with the approximate solution for large frameworks on pinned supports, derived in Chapter 5. Third, the exact solution for the stability problem of columns with both bending and shear stiffness will be presented. Finally, it will be shown how the results can be applied to the stability analysis of large frameworks.

6.4.1 Cantilever with Shear and Bending Deformation

In investigating the equilibrium of an elementary section of a cantilever developing both bending and shear deformation, subjected to uniformly distributed load along the axis of the bar (Fig. 6-12), the fourth-order differential equation of variable coefficients

$$y'''' - \frac{q\rho}{GA}(3y''' + zy'''') + \frac{q}{EI}(y' + zy'') = 0 \tag{6.25}$$

is obtained [Zalka, 1979], where the following notation is used:

- q - intensity of the applied vertical UDL,
- A - cross-sectional area of the column,
- I - second moment of area of the cross-section of the column,
- E - Young's modulus,
- G - modulus in shear,
- ρ - constant depending on the shape of the cross-section, (ρ=1.2 for rectangular cross-sections).

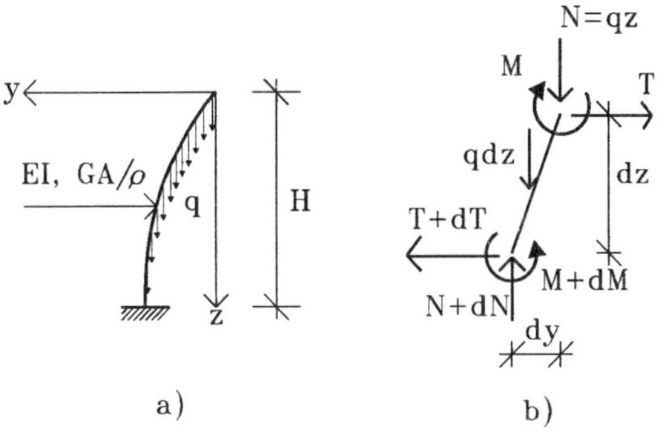

Fig. 6-12. Cantilever with bending and shear stiffness
a) cantilever and b) its elementary section

The boundary conditions for this equation are as follows.

a) The horizontal translation equals zero at the top of the column:

$$y(0) = 0\ . \tag{6.26a}$$

b) The tangent to the column at the bottom of the column is vertical:

$$y'(H)\left(1 - \frac{q\rho H}{GA}\right) = 0 \tag{6.26b}$$

c) The bending moments vanish at the top of the column:

$$y''(0) - \frac{\rho q}{GA} y'(0) = 0 \ . \qquad (6.26c)$$

d) The shear forces vanish at the top of the column:

$$y'''(0) - \frac{2 q \rho}{GA} y''(0) = 0 \ . \qquad (6.26d)$$

The order of differential equation (6.25) can be reduced. After integrating the equation once and making use of boundary condition (6.26d), we arrive at

$$y''' - \frac{q \rho}{GA}(zy''' + 2y'') + \frac{q}{EI} zy' = 0 \ . \qquad (6.27)$$

Boundary conditions (6.26a/b/c) belong to this equation.

Both the approximate and the exact solution to this equation are available [Zalka, 1979]. Nobody cares about the approximate solution under normal circumstances when the exact solution is available but in this case it has something to offer: The effect of the different stiffness characteristics is made clear and also the analogy with large frameworks is shown while the approximate procedure is produced.

Another reason for looking into both the approximate and exact solutions is that the eigenvalue problem is unique in a way: Not only the differential equation but also the boundary conditions contain the eigenvalue. To have the eigenvalue in the governing differential equation as well as in the boundary conditions makes the solution of the problem very difficult indeed in many cases. By showing the way in which this problem can be addressed, a method of general validity is offered, applicable to similar problems as well.

6.4.2 Approximate Solution

A simple approximate solution is available by the application of the Föppl-Papkovics theorem [Chapter 8]. Assume first that the column has infinitely great shear stiffness and only develops bending deformation. With $GA/\rho \to \infty$, equation (6.27) assumes the form

$$y''' + \frac{q}{EI} zy' = 0 \ . \qquad (6.28)$$

The solution to this equation results in the critical load of a cantilever subjected to uniformly distributed vertical load, having infinitely great shear stiffness:

$$N_{cr}^{b} = \frac{7.837EI}{H^2} \ . \qquad (6.29)$$

The characteristic bending type deformation is shown in Fig. 6-13/a.

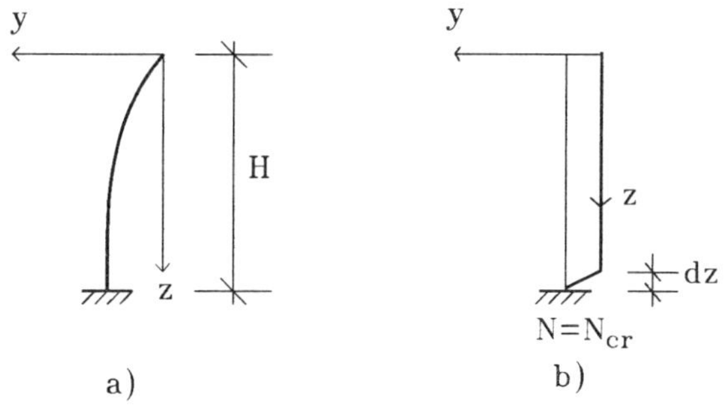

Fig. 6-13. a) Bending, b) shear type deformation of the column

If the bending stiffness of the column is infinitely great and it only undergoes shear deformation, equation (6.27) assumes the form

$$\left(1 - \frac{q\,z\rho}{GA}\right)y''' - \frac{2\,q\rho}{GA}\,y'' = 0\;. \tag{6.30}$$

From Eqn (6.30), after making use of the corresponding boundary conditions and carrying out some rearrangement, the equation

$$y'\left(N(z) - \frac{GA}{\rho}\right) = 0 \tag{6.31}$$

is obtained, where $N(z)=qz$. The boundary condition for this equation is defined by Eqn (6.26a). It is interesting to note that Eqn (5.38) in Section 5.3.2.1 assumes the same form. In carrying out the same procedure, we now obtain the eigenvalue of the problem, i.e., the critical load of a column having infinitely great bending stiffness, as

$$N^{s}_{cr} = \frac{GA}{\rho}\;. \tag{6.32}$$

The deflection "curve" is again unique: Straight, vertical axis with a sudden translation at the bottom of the column. Figure 6-13/b shows the corresponding shear type deformation.

The Föppl-Papkovics reciprocal theorem makes it possible to combine the two part critical loads and produce a lower bound to the resultant critical load. From the reciprocal addition

$$\frac{1}{N_{cr}} = \frac{1}{N^{b}_{cr}} + \frac{1}{N^{s}_{cr}} \tag{6.33}$$

the formula

$$N_{cr} = \frac{N^s_{cr}}{1 + \dfrac{N^s_{cr}}{N^b_{cr}}} . \tag{6.34}$$

is obtained.

This approximate formula is identical in form to formula (5.77) suggested in Section 5.4.3 for the approximate stability analysis of large frameworks on pinned supports, subjected to UDL at floor levels. Identity in form goes with identity in matter. According to the results, the approximate critical load is obtained by the reciprocal combination of two factors.

The resistance in shear is characterized by

$$K = K_b \frac{K_c}{K_c + K_b} \tag{6.35}$$

with frameworks on pinned supports (Eqn 5.54), and by

$$N^s_{cr} = \frac{GA}{\rho} \tag{6.36}$$

with columns with both bending and shear stiffnesses (Eqn 6.32). Both K and N^s_{cr} represent the shear stiffness of the corresponding structures.

The resistance in bending is characterized by

$$N_g = \frac{7.837 E_c I_g}{H^2} \tag{6.37}$$

with frameworks on pinned supports (Eqn 5.74), and by

$$N^b_{cr} = \frac{7.837 EI}{H^2} \tag{6.38}$$

with columns with both bending and shear stiffnesses (Eqn 6.29). Formulae (6.37) and (6.38) are identical in form, with $E_c I_g$ and EI representing the bending stiffness of the corresponding structures.

The above formulae clearly show the analogy between frameworks on pinned supports and columns with both shear and bending stiffnesses. As the analogy is established, the exact solution will now be presented.

6.4.3 Exact Solution

Having introduced the non-dimensional quantities

$$\lambda = \frac{qH\rho}{GA} = \frac{N_{cr}}{N^s_{cr}} \tag{6.39}$$

and

$$\bar{\beta} = \frac{GAH^2}{7.837EI\rho} = \frac{N^s_{cr}}{N^b_{cr}}\,, \tag{6.40}$$

equation (6.27) can be rewritten as

$$y''' - \frac{\lambda}{H}(2y'' + zy''') + \frac{7.837\lambda\bar{\beta}}{H^3}zy' = 0\,. \tag{6.41}$$

In Eqs (6.39), (6.40) and (6.41),

N_{cr} - is the critical load,

N^s_{cr} - is the critical load associated with the shear stiffness,

N^b_{cr} - is the critical load associated with the bending stiffness.

Equation (6.41) can be solved by the power series method and then the eigenvalue of the problem can be produced for different stiffness ratios. Table 6-4 and Fig. 6-14 present values for load parameter λ in the function of $\bar{\beta}$. With these values the critical load of columns having both shear and bending stiffness is obtained from Eqn (6.39):

$$N_{cr} = \lambda N^s_{cr}\,. \tag{6.42}$$

The above formulae offer a solution for structures subjected to uniformly distributed normal load along the vertical axis. A closed form solution is also available for structures subjected to a load of arbitrary distribution [Hegedüs and Kollár, 1984a].

Table 6-4 also contains the deviation (Δ) of the Föppl-Papkovics solution from the continuum solution. According to the data, the Föppl-Papkovics additive formula is always conservative. The maximum deviation from the exact value is +29 per cent.

If the column is subjected to a single concentrated force at the top, the exact critical load is obtained from

$$F_{cr} = \frac{F^s_{cr}}{1 + \dfrac{F^s_{cr}}{F^b_{cr}}} . \tag{6.43}$$

Table 6-4. Values for the critical load parameter λ
λ : Continuum solution, λ_F : Föppl-Papkovics, Δ : Föppl's error

$\bar{\beta}$	λ_F	λ	Δ	$\bar{\beta}$	λ_F	λ	Δ	$\bar{\beta}$	λ_F	λ	Δ
0.0	1.000	1.000	0	2.5	0.286	0.334	15	5	0.1667	0.1824	9
0.1	0.909	1.000	9	2.6	0.278	0.323	14	6	0.1429	0.1543	7
0.2	0.833	1.000	17	2.7	0.270	0.313	14	7	0.1250	0.1337	7
0.3	0.769	1.000	23	2.8	0.263	0.304	13	8	0.1111	0.1179	6
0.4	0.714	0.997	28	2.9	0.256	0.295	13	9	0.1000	0.1055	5
0.5	0.667	0.933	29	3.0	0.250	0.287	13	10	0.0909	0.0954	5
0.6	0.625	0.866	28	3.1	0.244	0.279	12	12	0.0769	0.0802	4
0.7	0.588	0.805	27	3.2	0.238	0.271	12	14	0.0667	0.0691	4
0.8	0.556	0.750	26	3.3	0.233	0.264	12	16	0.0588	0.0607	3
0.9	0.526	0.701	25	3.4	0.227	0.257	12	18	0.0526	0.0541	3
1.0	0.500	0.658	24	3.5	0.222	0.251	11	20	0.0476	0.0488	3
1.1	0.476	0.619	23	3.6	0.217	0.245	11	30	0.0323	0.0328	2
1.2	0.455	0.584	22	3.7	0.213	0.239	11	40	0.0244	0.0247	1
1.3	0.435	0.553	21	3.8	0.208	0.233	11	50	0.0196	0.0198	1
1.4	0.417	0.524	21	3.9	0.204	0.228	11	60	0.0164	0.0165	1
1.5	0.400	0.499	20	4.0	0.200	0.223	10	70	0.0141	0.0142	1
1.6	0.385	0.476	19	4.1	0.196	0.218	10	80	0.0123	0.0124	1
1.7	0.370	0.454	18	4.2	0.192	0.213	10	90	0.0110	0.0111	1
1.8	0.357	0.435	18	4.3	0.189	0.209	10	100	0.00990	0.00995	1
1.9	0.345	0.417	17	4.4	0.185	0.205	10	150	0.00662	0.00665	0
2.0	0.333	0.401	17	4.5	0.182	0.201	9	200	0.00498	0.00499	0
2.1	0.323	0.385	16	4.6	0.179	0.197	9	300	0.00332	0.00333	0
2.2	0.313	0.371	16	4.7	0.175	0.193	9	400	0.00249	0.00250	0
2.3	0.303	0.358	15	4.8	0.172	0.189	9	500	0.00200	0.00200	0
2.4	0.294	0.346	15	4.9	0.170	0.186	9	1000	0.00100	0.00100	0

In formula (6-43)

$$F^s_{cr} = \frac{GA}{\rho} \tag{6.44}$$

is the critical load associated with the shear stiffness of the column and

$$F^b_{cr} = \frac{\pi^2 EI}{4H^2} \tag{6.45}$$

is the critical load associated with the bending stiffness of the column.

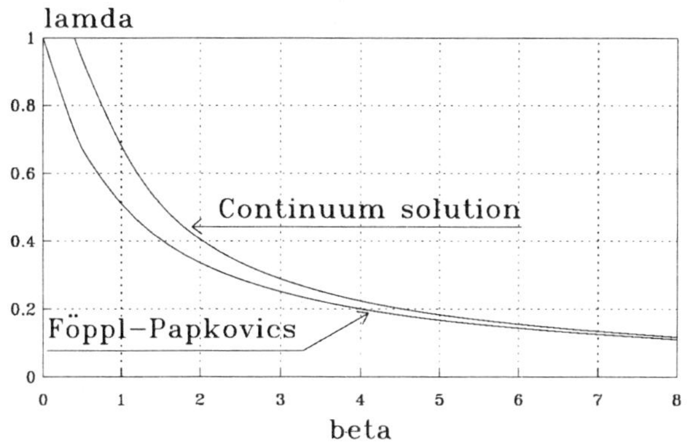

Fig. 6-14. Values for critical load parameter λ

6.4.4 Application to Frameworks

An analogy between the two problems has been established and the exact solution to the cantilever with bending and shear stiffnesses has been produced. Application to large frameworks will be presented in this Section.

However, there is also some difference between the two cases, which has to be considered first. When the governing equation of the problem (Eqn 6.25) was established and formula (6.42) for the critical load was produced, only the term "bending stiffness" was used. To be precise, it should have been necessary to differentiate between local and global bending stiffness. This difference is well known in the theory of sandwich columns consisting of layers (faces) and connecting media [Allen, 1969]. Figure 6-15 shows the model of a sandwich column whose shear stiffness is defined by S. The layers are called thick if the second moment of area of the layers with respect to both their own centroidal axis (l) and also to the centroidal axis of the layers (g) is taken into account. If the own second moment of area of the layers (I_l) is neglected, the layers are called thin. The column investigated above was in fact a sandwich column and the stiffness referred to as bending stiffness was in fact the global bending stiffness. It follows that formula (6.42) can only be applied to frameworks with no or negligible local bending stiffness. Bearing this in mind, the analysis can be carried out as follows.

With one-bay or multi-bay frameworks, the equivalent two-layer or multi-layer sandwich column (Fig. 6-16) is established first. Shear stiffness K (Eqn 6.35) is then computed, depending on the support system and the nature of the additional local sway of the framework. With the critical load associated with global bending N^b_{cr} (Eqn 6.37), parameter $\bar{\beta}$

is obtained on the basis of formula (6.40) as

$$\bar{\beta} = \frac{N^s_{cr}}{N^b_{cr}} = \frac{K}{N_g} . \tag{6.46}$$

Making use of Table 6-4 or Fig. 6-14, formula (6.42) gives the critical load for frameworks subjected to UDL on the beams:

$$N_{cr} = \lambda K . \tag{6.47}$$

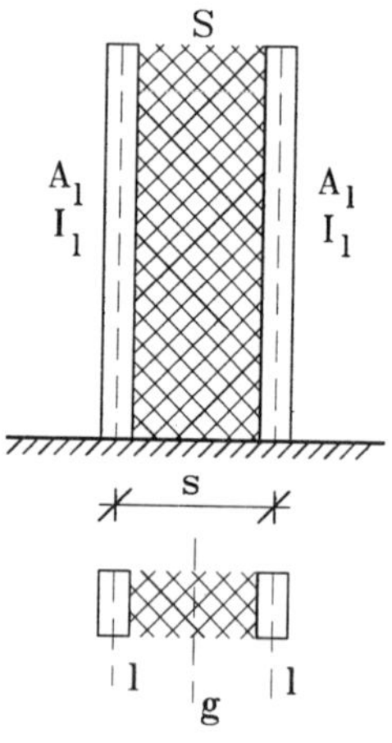

Fig. 6-15. Sandwich column with two faces and connecting media

If the critical load for frameworks subjected to concentrated forces at top floor level is needed, formula (6.43) should be used. The shear stiffness in formula (6.43) is expressed by

$$F^s_{cr} = K = K_b \frac{K_c}{K_c + K_b} \tag{6.48}$$

and the global bending stiffness is obtained from

$$F^b_{cr} = F_g = \frac{\pi^2 EI_g}{4H^2} . \tag{6.49}$$

The procedure is surprisingly simple and offers reliable critical loads for large frameworks on pinned supports, whose columns are connected by beams at ground floor level. Because of the pinned supports, the columns of the framework cannot activate their local bending stiffness so that the thin-layer sandwich column model is acceptable for the framework. However, there is some discrepancy in the model. As it was pointed out in Section 5.3.2.1, the local bending stiffness does influence the critical load, although only in an indirect way.

Because of the absence of the local bending stiffness in the formulae, these formulae cannot cover frameworks on fixed supports. They can be used as rough estimates but real analogy only exists if the local bending stiffness is introduced into the governing differential equation of the sandwich column. This procedure leads to the establishment of the equation of sandwich columns with thick faces. The procedure requires lengthy mathematical derivations and is therefore beyond the scope of this book. However, the detailed analysis is available in [Hegedüs and Kollár, 1984b; Kollár, 1988].

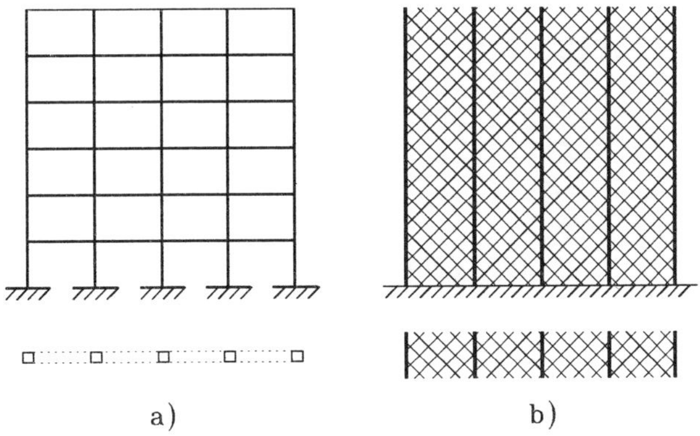

Fig. 6-16. (a) Multi-bay framework, (b) multi-layer sandwich column

The following numerical example shows how to use the formulae for the rapid stability check of large frameworks.

6.4.5 Numerical Example

The formulae derived in Section 6.4 will now be applied to framework PFG4 shown in Fig. 5-33 in Section 5.6.2.1. The four-bay, twenty-storey framework is proportional. The critical load for the framework subjected to UDL on the beams will be determined.

The following characteristics are needed for the calculation.

Stiffness of the columns (Eqn 5.53):

$$K_c = \sum_1^n \frac{\pi^2 E_c I_{c,i}}{h^2} = \frac{\pi^2 \times 3\times 10^4 \times 8.533\times 10^{-3}}{3^2} = 280.74 \text{ MN} .$$

Stiffness of the beams (Eqn 5.9):

$$K_b = 2\sum_1^{n-1} \frac{6E_b I_{b,i}}{l_i h} = \frac{2\times 4\times 6\times 3\times 10^4\times 1.143\times 10^{-2}}{3\text{x}3} = 1829.33 \text{ MN} .$$

Shear stiffness (Eqn 6.35):

$$K = K_b \frac{K_c}{K_c + K_c} = \frac{280.74}{280.74 + 1829.33} = 243.38 \text{ MN} .$$

Second moment of area of the cross-sections of the columns for the global bending stiffness (Eqn 5.20):

$$I_g = \sum_1^n A_{c,i} t_i^2 = 2(0.127\times 6^2 + 0.16\times 3^2) = 12.024 \text{ m}^4.$$

Global bending critical load (6.37):

$$N_g = \frac{7.837 E_c I_g}{H^2} = \frac{7.837\times 3\times 10^4\times 12.024}{60^2} = 785.27 \text{ MN} .$$

Stiffness ratio (Eqn 6.46):

$$\bar{\beta} = \frac{N^s_{cr}}{N^b_{cr}} = \frac{K}{N_g} = \frac{243.38}{785.27} = 0.31 .$$

With stiffness ratio $\bar{\beta}$, after linear interpolation the critical load parameter is obtained from Table 6-4:

$$\lambda = 0.999 .$$

Formula (6.47) finally gives the critical load:

$$N_{cr} = \lambda K = 0.999 \times 243.38 = 243.14 \text{ MN} .$$

The exact solution for the critical load is

$$N_{cr} = 256.0 \text{ MN} .$$

The deviation from the exact critical load is 5 per cent.

When the framework is subjected to concentrated forces at top floor level, formula (6.43) yields the critical load. The part critical load which is associated with the shear stiffness has already been computed:

$$F^{s}_{cr} = K = 243.38 \text{ MN} .$$

The part critical load associated with the global bending stiffness is obtained from formula (6.49):

$$F^{b}_{cr} = F_{g} = \frac{\pi^2 EI_g}{4H^2} = \frac{\pi^2 \times 3 \times 10^4 \times 12.024}{4 \times 60^2} = 247.23 \text{ MN} .$$

With the above quantities, the critical load assumes the value

$$F_{cr} = \frac{1}{1 + \dfrac{243.38}{247.23}} = 122.7 \text{ MN} .$$

The exact critical load is

$$F_{cr} = 127.5 \text{ MN} .$$

The error made by the application of the sandwich analogy is 4 per cent.

6.5 SUPPLEMENTARY REMARKS

Without giving a detailed analysis, we only show how two practical cases can be dealt with in the continuum stability analysis of frameworks. The effect of semi-rigid beam-column joints on the critical load of frameworks can be taken into account by slightly modifying the formula of the beam stiffness. The formulae derived for the elastic critical load also make it possible to carry out a simple elasto-plastic stability analysis.

6.5.1 Semi-rigid Joints

Semi-rigid beam-column joints are easy to be taken into account. If the joint between the beams and the columns is flexible, then the fixing effect of the beams on the columns is reduced, according to the flexibility of the joint. This effectively means a reduction of the stiffness of the beams. Introducing factor m into the expression of the stiffness of the beams, the semi-rigid nature of the joints is approximately allowed for:

$$\bar{k}_b = mk_b \ . \tag{6.50}$$

The value of factor m varies from zero to one. Value m=1 represents the fixed case and m=0 stands for the pinned joint. With the modified beam stiffness (6.50), all the continuum formulae can be used. With factor m varying from joint to joint, formula (6.50) even makes it possible to take into consideration that the flexibility of the joints may vary horizontally from joint to joint.

Another approach to the problem is the application of Dunkerley's formula in the form

$$\bar{k}_b = \left(\frac{1}{k_b} + \frac{1}{k_j} \right)^{-1} \tag{6.51}$$

where k_j is the joint stiffness. In formulae (6.50) and (6.51) k_b is the stiffness of the beam defined by formulae (5.5) and (5.9).

Values for parameter m and joint stiffness k_j are to be found according to the geometrical characteristics of the joint and the relevant Code of Practice. Howlett [1981] and Jones et al [1979] are valuable sources for further reference.

6.5.2 Elasto-plastic Analysis

It is often not enough to carry out an elastic stability analysis because the structure responds in the elasto-plastic region. The exact second order elasto-plastic stability analysis is a formidable task even for small frameworks. However, the well-known Merchant-Rankine formula [Chapter 8]

$$F_{cr}^{ep} = \left(\frac{1}{F_{cr}^{e}} + \frac{1}{F_{cr}^{p}} \right)^{-1} \tag{6.52}$$

offers a simple estimate for the elasto-plastic critical load. In formula (6.52)

F_{cr}^{e} is the elastic critical load,

F_{cr}^{p} is the rigid-plastic collapse load

and

F_{cr}^{ep} is the elasto-plastic critical load.

The beauty of formula (6.52) is in its simplicity. For steel unbraced frameworks, the rigid-plastic collapse load is relatively easy to produce and the simple formulae presented in Chapters 5 and 6 can be used for the elastic critical load. According to accuracy analyses [Anderson and Lok, 1983], the Merchant-Rankine formula is conservative when realistic horizontal and vertical loads are applied. Modified versions of the Merchant-Rankine formula are also available [Wood, 1974; Anderson and Lok, 1983], to improve the accuracy and to include the effects of strain-hardening and cladding.

It is interesting that the Merchant-Rankine formula is not a generalized Dunkerley formula as its form indicates. It is still an open question, whether or not the approximation made by its application is conservative [Tarnai, 1981], According to Horne's research, it is conservative with frameworks, as a rule [Horne, 1963 and 1981].

6.5.3 Generalization

The approach presented in Chapters 5 and 6 to the stability analysis of frameworks and coupled shear walls was to create a simple model, a substitute column, which was complex enough to take into consideration two of the most important characteristics; the global shear deformation and the local bending deformation of the columns. The governing differential equation of the problem was then established and solved. Finally, relying on practical considerations and analysing the deformation of the structures under realistic load systems, the remaining characteristics were approximately taken into consideration. This approach made it possible to deal with and to show the effect of the important stiffness characteristics one by one.

Another way to address the problem is to establish a very sophisticated continuum model which can allow for most of the important characteristics. While the former approach offers more to structural designers, the letter one has great theoretical importance. The simple continuum model was discussed in Chapters 5 and 6 and details of more complex generalized bar models are available in publications by Hegedüs and Kollár [1988a and 1988b] and Kollár [1986].

7

SPATIAL ANALYSIS OF REGULAR STRUCTURES UNDER NORMAL LOADING

The load bearing system of buildings often consists of frameworks, shear walls, coupled shear walls and cores (vertical load bearing elements) and floor slabs (horizontal load bearing elements). The floor slabs are responsible for transmitting the applied vertical loads to the vertical load bearing elements and for distributing the horizontal loads among the vertical load bearing elements. The lateral stiffness of a building is basically provided by the vertical load bearing elements. Such structures, subjected to vertical loads, can develop three kinds of deformation: Buckling in the two principal planes of inertia and pure torsional buckling. Apart from some special symmetric arrangements, the three modes combine and the building undergoes torsional buckling.

Even the exact stress analysis of these structures is rather complicated, partly because of the interaction between the horizontal and vertical systems and among the elements of the horizontal and vertical systems themselves and partly because of the great number of elements to be involved in the analysis. The finite element method has proved to be efficient for the stress analysis [Zienkiewitz, Parekh and Teply, 1971]. As far as the stability analysis of large structures is concerned, however, the preparation of the input data and the actual computation require relatively much more time.

The method of sub-structures used for the stress analysis by Harman and Walker [1975] and Nair [1975] offers another powerful possibility. Relying on this method, a large structure can be divided into smaller parts either in the horizontal or in the vertical direction, increasing efficiency and decreasing computation time.

In investigating the stresses of a spatial system of coupled shear walls, Horacek [1970, 1971, 1974] applies the principle of minimum strain energy and considers the sums of the shear forces in the lintels as unknown quantities. The three dimensional analysis is in this way simplified to the analysis of cantilevers. This method, however, cannot deal with cores of different shapes.

Yet another useful approach is offered by the theory of thin-walled structures. The application often results in simple, closed form solutions and this Chapter shows how the theory of thin-walled structures can be applied to the three dimensional stability analysis of buildings.

7.1 INTRODUCTION

The method to be used here is based on the assumption that the vertical load bearing system of the building consists of vertical cantilevers of thin-walled open or closed cross-section. These cantilevers are made to work together by the floor slabs. The method, which introduces a substitute column, has widely been used for the stress analysis of tall buildings [Taranath, 1968 and 1988; Beck and Shäfer, 1969; Szmodits, 1975].

Most methods available for the stability analysis rely on the method of thin-walled structures [Podolski, 1970; Glück and Gellert, 1972; Yarimci, 1972; Danay, Gellert and Glück, 1975; Rosman, 1980]. According to investigations into the behaviour of large structures, under certain circumstances, the overall stability analysis of buildings can be traced back to the torsional buckling analysis of a single cantilever of thin-walled section, subjected to uniformly distributed vertical load. The biggest challenge through the procedure is the solution of the eigenvalue problem of the differential equation of variable coefficient of pure torsional buckling. Instead of solving this problem, the authors of approximate procedures often introduce different simplifying assumptions concerning the torsional or warping characteristics of the substitute cantilever. Relying on Southwell's additive theorem and the Föppl-Papkovics reciprocal additive theorem, Kollár [1977] suggests a simple, expressive formula for the approximate critical load of buildings with a single core, subjected to torsional buckling. Rosman [1980] assumes that the flexural and torsional buckling modes can be characterized by cosine functions and arrives at a simple cubic equation, the smallest root of which results in the critical load. Both methods produce a lower bound to the critical load of pure torsional buckling and consequently the overall critical load they present is conservative. Akesson's [1980] research on rack structures results in simple formulae for the critical load of structures with negligible torsional stiffness.

In the following sections, simple formulae will be given for the stiffness characteristics of the substitute column of regular building structures, the governing differential equations of torsional buckling will be established and solved and finally closed form solutions will be presented for the elastic critical load.

Only regular structures are considered. Structures are defined regular, if

a) the floor slabs are at a regular distance from each other, in other words, if the storey height is constant and

b) the arrangement and the cross-sections of the vertical load bearing elements do not vary along the height of the building.

In addition to the "regularity requirements" a) and b) above, the following assumptions must be met:

1) The material of the structural elements is homogeneous, isotropic and obeys Hooke's law.

2) The stiffness of the floor slabs of the building is infinitely great in their plane and infinitely small perpendicular to their plane.

3) The elements of the support system of the building are rigidly clamped shear walls and cores, i.e., they are sway structures.

4) The geometrically perfect structures develop small deformations.

5) The position of the shear centre only depends on geometrical characteristics.

7.2 SUBSTITUTE CORE

With cores subjected to torsional buckling, the location of the shear centre and the axis passing through the shear centre play an important role. The location of the shear centre is best determined in the co-ordinate system $\bar{x}$-$\bar{y}$, whose origin is in the left up corner of the ground plan of the building with the co-ordinate axes bordering the ground plan of size *B* and *L* (Fig. 7-1).

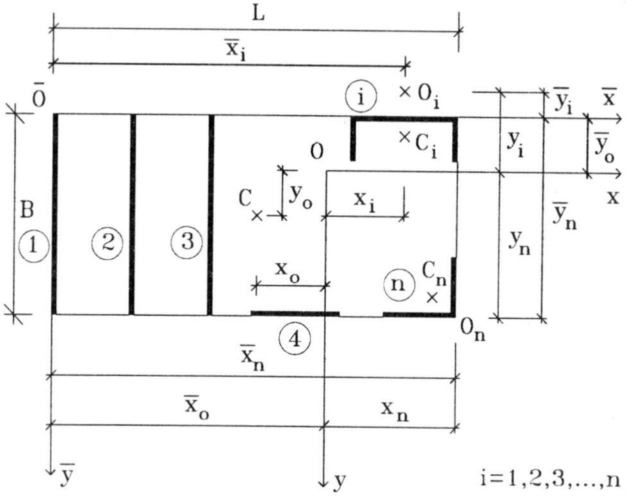

Fig. 7-1. Ground plan with support system

The co-ordinates of shear centre *O* in this co-ordinate system are obtained from the following formulae [Beck and Shäfer, 1969]:

$$\bar{x}_o = \frac{I_{xy}(\Sigma I_{y,i}\bar{y}_i - \Sigma I_{xy,i}\bar{x}_i) - I_y(\Sigma I_{xy,i}\bar{y}_i - \Sigma I_{x,i}\bar{x}_i)}{I_x I_y - I_{xy}^2} \tag{7.1}$$

and

$$\bar{y}_o = \frac{I_x(\Sigma I_{y,i}\bar{y}_i - \Sigma I_{xy,i}\bar{x}_i) - I_{xy}(\Sigma I_{xy,i}\bar{y}_i - \Sigma I_{x,i}\bar{x}_i)}{I_x I_y - I_{xy}^2} \tag{7.2}$$

where $I_{x,i}$, $I_{y,i}$ and $I_{xy,i}$ represent the second moments of area and the product of inertia of the i-th element of the support system with respect to axes x and y, respectively, in the local centroidal co-ordinate system of the elements. Quantities

$$I_x = \Sigma I_{x,i}, \qquad I_y = \Sigma I_{y,i}$$

and

$$I_{xy} = \Sigma I_{xy,i}$$

are the sums of the second moments of area and the product of inertia of the elements of the support system, where i=1,...,n. Co-ordinates $\bar{x}_i$ and $\bar{y}_i$ represent the location of the shear centre of the individual elements (cores, shear walls) in the co-ordinate system $\bar{x}$ and $\bar{y}$. The number of support elements is n.

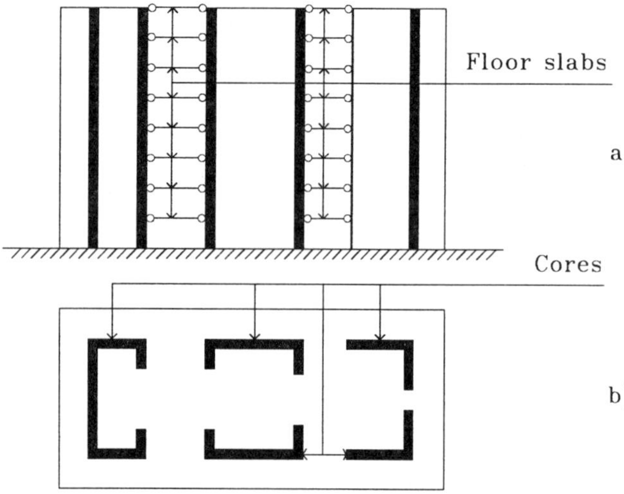

Fig. 7-2. Model for flexural buckling: a) elevation, b) plan view

Knowing the location of the shear centre and the geometrical characteristics of the cores, a substitute cantilever can be created, by incorporating the individual cores into a single cantilever. The shear centre of this cantilever coincides with the shear centre of the support system of the building. The stiffnesses of the substitute cantilever are special sums of the individual cores, computed by the following principles. For the bending and pure torsional stiffnesses, the corresponding values of individual cores have to be added up:

$$EI_x = E\Sigma I_{x,i}, \tag{7.3}$$

$$EI_y = E\Sigma I_{y,i}, \tag{7.4}$$

$$EI_{xy} = E\Sigma I_{xy,i}, \tag{7.5}$$

$$GJ = G\Sigma J_i, \tag{7.6}$$

where E is the modulus of elasticity, G is the modulus of elasticity in shear and J_i is the St Venant torsion constant of the i-th core. The establishment of formulae (7.3)-(7.6) follows from the fact that the stiffness of the floor slabs of the building is infinitely great in their plane and infinitely small perpendicular to their plane (Assumption 2 in Section 7.1). The infinitely great in-plane stiffness of the floor slabs makes the cores work together in the horizontal direction, resulting in identical deflections. However, the infinitely small stiffness of the floor slabs perpendicularly to their plan cannot make them work together in the vertical direction so that the individual cores cannot act as the longitudinal fibres of a bar and do not develop axial deformation from the point of view of the whole system. It follows that, as far as bending is concerned, the floor slabs act like pinned bars linking the cores (Fig. 7-2). The warping stiffness of the substitute column is obtained from the following formula:

$$EI_\omega = E\sum\left\{I_{\omega,i} + I_{x,i}x_i^2 + I_{y,i}y_i^2 - 2I_{xy,i}x_iy_i\right\}. \tag{7.7}$$

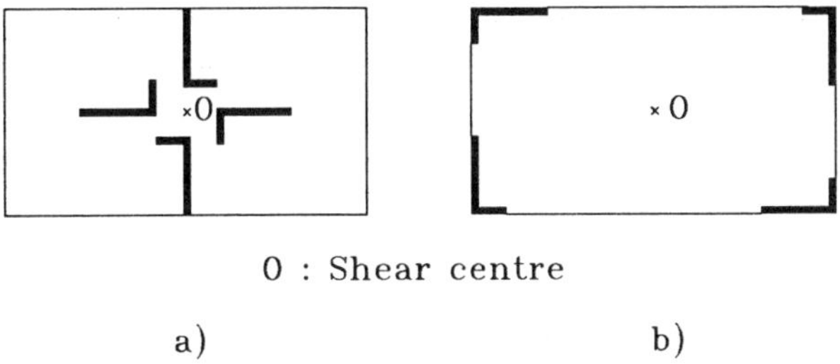

Fig. 7-3. a) Inefficient and b) good arrangement of the support system

Co-ordinates x_i and y_i in this formula belong to the i-th core in the co-ordinate system x-y, whose origin coincides with the shear centre of the support system and whose x and y axes are parallel with the corresponding sides of the ground plan of the building (Fig. 7-1). Formula (7.7) needs some explanation. The first term in Eqn (7.7) shows that all the cores may develop warping deformation. The remaining terms in the formula are responsible for the fact that the substitute cantilever as a whole can also develop warping since the floor slabs (the "cross-sections" of the substitute cantilever) are infinitely flexible perpendicularly to their plane and do not remain plane. This second part expressively shows that the individual cores, apart from their own warping stiffnesses, contribute to the overall warping stiffness of the building in proportion to their bending stiffnesses and their distance to the shear centre [Brandt, Shäfer and Reeh, 1975].

The distance of a core from the shear centre plays an important role, as seen in formula (7.7), indicating the importance of the geometrical arrangement of the cores. Examples for good and inefficient arrangement, using the same amount of material, are presented in Fig. 7-3.

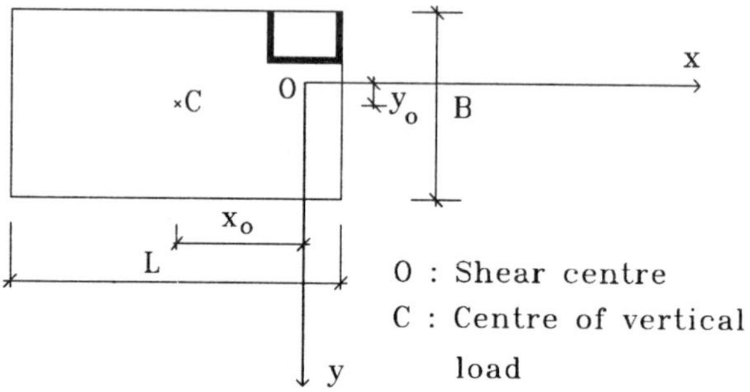

Fig. 7-4. Rectangular ground plan

Apart from the above stiffnesses, the radius of gyration is also needed. Unlike with the bending, torsional and warping stiffnesses, the radius of gyration is only determined by the size of the ground plan of the building. This follows from the derivation of the differential equation of pure torsional buckling. The derivation of the equation is based on the analysis of the behaviour of the longitudinal fibres of the cantilever during buckling [Zalka, 1980a]. The external vertical loads act along these fibres causing torsion. The cantilever balances this effect by its torsional and warping stiffnesses. With buildings, the external loads act on the floors and they are transmitted by the columns, shear walls and cores (longitudinal fibres). It follows therefore that, given a certain area and arrangement of support elements, the larger the ground plan of a building, the smaller the value of the critical load for pure torsional buckling because the moment arms of the forces which cause torsion are great and so the value of the external torque is great as well. On the other hand, a good arrangement of the shear walls and cores can result in considerable warping resistance (Fig. 7-3/b). The expression for the radius of gyration for a rectangular ground plan is

$$i_p = \sqrt{\frac{(L^2 + B^2)}{12} + t^2} \tag{7.8}$$

where L and B are the size of the ground plan and

$$t = \sqrt{(x_o^2 + y_o^2)}$$

is the distance between shear centre O and the centre of the resultant of the vertical loads C (Fig. 7-4). Note that the origin of the coordinate system and the shear centre coincide.

There is an interesting - and important - difference between the radius of gyration (7.8) and the stiffness characteristics of the substitute column, defined by Eqs (7.3)...(7.6). The radius of gyration is related to the size of the ground plan of the building and not to the size of the cores. The reason for this fact is clearly seen from the derivation of the governing differential equation of torsional buckling (Chapter 8).

Having established the geometrical and stiffness characteristics of the substitute cantilever, the next step is the establishment and solution of the system of differential equations of torsional buckling for different load cases.

7.3 UNIFORMLY DISTRIBUTED FLOOR LOADS

When the building is subjected to uniformly distributed load on the floors, the model for the analysis is a cantilever (the substitute cantilever) subjected to concentrated forces at floor levels. In distributing these concentrated forces between the floors, a continuum model is obtained (Fig. 7-5). This model will be investigated in this section. Buildings subjected to concentrated forces at top floor level will be dealt with in Section 7.4.

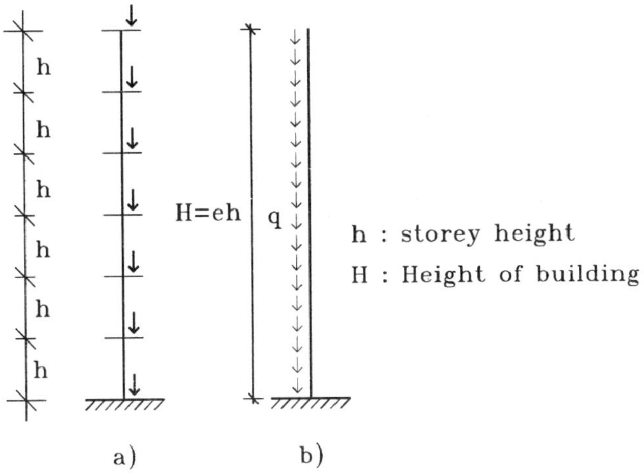

Fig. 7-5. Continuum modelling of the vertical load on the floors

7.3.1 The Governing Differential Equations

Consider the column with arbitrary cross-section, having a free upper and a fixed lower end (Fig. 7-6). This column is the substitute cantilever of the building, with the geometrical and stiffness characteristics given in Section 7.2.

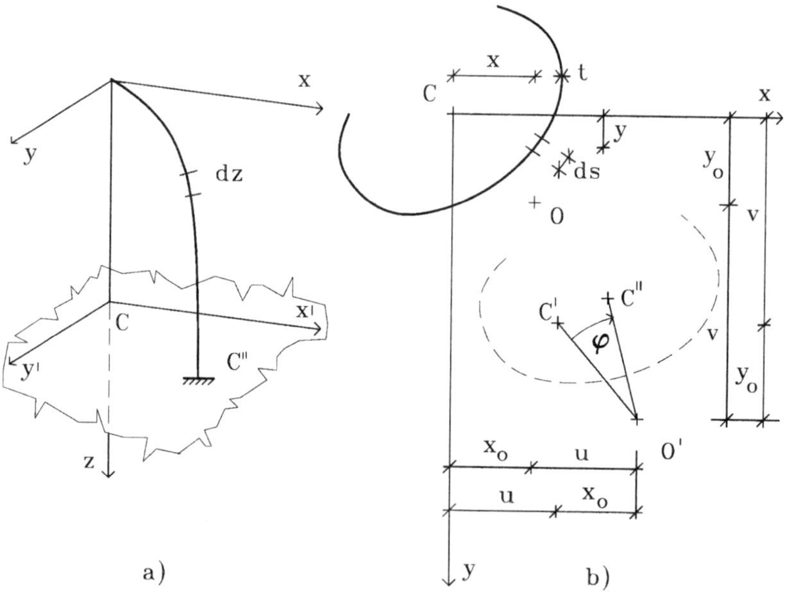

Fig. 7-6. Substitute column: a) 3D view, and b) cross-section

The cantilever is subjected to a uniformly distributed load of intensity q along the height. Set the co-ordinate system in such a way that axis z coincides with the centroidal axis of the bar and axes x and y are the principal axes of inertia of the free upper cross-section of the bar. Axes x and y are fixed to the upper cross-section and therefore they remain principal axes during buckling as well (Fig 7-6/a). Point C denotes the centroid and x_o and y_o are the co-ordinates of shear centre O. During buckling the cross-sections of the bar undergo translation and rotation. The translations are defined by deflections u and v in direction x and y, respectively, of the shear centre. During the translation of the cross-section at distance z from the origin, point O moves to point O' and point C to point C'. The rotation of the cross-section about the shear centre is denoted by angle φ and the final position of the centroid is C'' (Fig. 7-6/b). Examining the equilibrium of an elementary section of length dz of the shear centre axis, the following system of differential equations is obtained:

$$EI_y u'''' + \left(N(u' + y_o \varphi')\right)' = 0, \tag{7.9}$$

$$EI_x v'''' + \left(N(v' - x_o \varphi')\right)' = 0, \tag{7.10}$$

$$EI_\omega \varphi'''' - GJ\varphi'' + \left(N\frac{I_o}{A}\varphi'\right)' - \left(N(x_o v' - y_o u')\right)' = 0. \tag{7.11}$$

Eqs (7.9) and (7.10) are the governing differential equations of bending and their derivation is available in most structural engineering textbooks. Eqn (7.11) characterizes the torsion of the bar and, because of its importance, the derivation is given in Chapter 8.

In Eqs (7.9), (7.10) and (7.11) $N=qz$, which is the sum of the UDL at z, A is the area of the ground plan of the building and

$$I_o = I_x + I_y + A\left(x_o^2 + y_o^2\right)$$

is the polar moment of inertia of the ground plan about the shear centre.

Since axes x and y have been fixed on the upper cross-section of the bar, the boundary conditions for Eqs (7.9) and (7.10) are given as the following.

a) The deflection on the top of the bar equals zero in directions x and y:

$$u(0) = v(0) = 0. \tag{7.12}$$

b) At the bottom of the bar the slope of the deflection in directions x and y is parallel with axis z:

$$u'(H) = v'(H) = 0. \tag{7.13}$$

c) The bending moment equals zero at the top of the bar:

$$u''(0) = v''(0) = 0. \tag{7.14}$$

d) The shear force equals zero at the top of the bar:

$$u'''(0) = v'''(0) = 0. \tag{7.15}$$

Equation (7.11) allows for torsion. The boundary conditions accompanying Eqn (7.11) are given below.

a) Rotation at the top of the bar equals zero:

$$\varphi(0) = 0. \tag{7.16}$$

b) The bottom of the bar is fixed and no warping develops there:

$$\varphi'(\mathrm{H}) = (0). \tag{7.17}$$

c) No compressive forces act at the top cross-section of the bar so warping stresses cannot develop:

$$\varphi''(0) = (0). \tag{7.18}$$

d) The torsional moment at the top of the bar equals zero:

$$GJ\varphi'(0) - EI_{\omega}\varphi'''(0) = 0. \tag{7.18a}$$

Equations (7.9), (7.10) and (7.11) are three simultaneous differential equations for buckling by bending and torsion and can be used for the determination of the critical load. The angle of rotation φ appears in all three equations showing that in the general case torsional buckling and the bending of the bar in both principal axes of inertia occur simultaneously. In the case of columns with thin-walled open cross-section, the phenomenon of torsional-flexural buckling is very important because the critical load of pure torsional buckling can be considerably smaller than those of planar buckling and the critical load of the combined torsional-flexural buckling is small as well.

For the calculation of the critical load of a column with monosymmetric or unsymmetric cross-section, the value of the critical load for pure torsional buckling is needed. This can be obtained by the analysis of columns with doubly symmetric cross-section. This procedure will now be presented.

7.3.2 Doubly Symmetric Layout

In the special case when the cross-section has two axes of symmetry, the shear centre and the centroid coincide. If, in addition, the resultant of the vertical loads passes through the centroid, then

$$x_o = y_o = 0$$

holds and, after the reduction of the order of Eqs (7.9), (7.10) and (7.11), the governing differential equations of the problem become

$$EI_y u''' + qzu' = 0, \tag{7.19}$$

$$EI_x v''' + qzv' = 0, \tag{7.20}$$

$$EI_{\omega}\varphi''' - \left(GJ - qz\frac{I_o}{A}\right)\varphi' = 0. \tag{7.21}$$

Each of these equations only contains one unknown quantity and therefore they can be treated separately. It follows that pure torsional buckling is independent of flexural buckling in both principal planes of inertia. Accordingly, through the stability analysis of bars having two axes of symmetry, and buildings with doubly symmetric load and support system arrangement, the following possibilities have to be taken into account.

a) The centrally compressed structure develops bending in one of the two principal planes *xy* and *yz*, i.e., flexural buckling occurs. In this case the critical load can be obtained by solving the eigenvalue problems given by Eqs (7.19) and (7.20). Equations (7.12)...(7.14) give the boundary conditions. Equations of this type are traceable to Bessel equations. The use of Bessel functions results in the following formulae for the critical load [Timoshenko and Gere, 1961]:

$$N_{cr,x} = qH = \frac{7.837EI_y}{H^2}, \tag{7.22}$$

and

$$N_{cr,y} = qH = \frac{7.837EI_x}{H^2}. \tag{7.23}$$

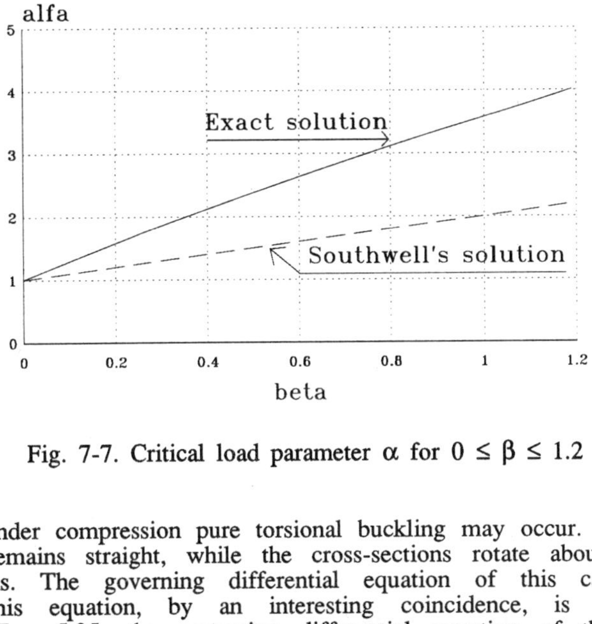

Fig. 7-7. Critical load parameter α for $0 \leq \beta \leq 1.2$

b) Under compression pure torsional buckling may occur. The axis of the bar remains straight, while the cross-sections rotate about the shear centre axis. The governing differential equation of this case is Eqn (7.21). This equation, by an interesting coincidence, is identical in form to Eqn 5.35, the governing differential equation of the substitute column of frameworks subjected to uniformly distributed vertical load along the height. The local bending stiffness of the columns and the global shear stiffness of the framework is now replaced by the warping stiffness and the torsional stiffness of the substitute cantilever, respectively. The solution derived in Chapter 5 can be directly used.

By so doing, the solution of the eigenvalue problem (7.21), i.e., the critical load for pure torsional buckling, is obtained as

$$N_{cr,\varphi} = \alpha N^{\omega}_{cr,\varphi}, \tag{7.24}$$

where α is the eigenvalue of the problem.

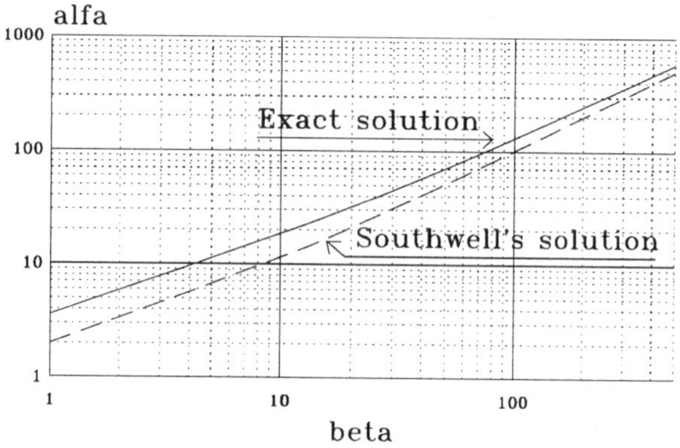

Fig. 7-8. Critical load parameter α for $1 \leq \beta \leq 316$

Values for parameter α are given in Table 7-1 and in Figs 7-7 and 7-8, in the function of the stiffness ratio

$$\beta = \frac{N^{t}_{cr,\varphi}}{N^{\omega}_{cr,\varphi}} . \tag{7.25}$$

In formula (7.25)

$$N^{t}_{cr,\omega} = \frac{1}{i_p^2} GJ \tag{7.26}$$

is the "part" critical load, associated with the torsional stiffness and

$$N^{\omega}_{cr,\varphi} = \frac{7.837 EI_{\omega}}{i_p^2 H^2} \tag{7.27}$$

is the "part" critical load, associated with the warping stiffness.

In the above formulae, the radius of gyration was introduced as

$$i_p = \sqrt{\frac{I_o}{A}} .$$

Southwell's additive theorem, often used in structural engineering practice, results in

$$N_{cr,\varphi} = \frac{1}{i^2}\left(\frac{7.837EI_{\omega}}{H^2} + GJ\right) \tag{7.28}$$

for the critical load of pure torsional buckling.

Table 7-1. Critical load parameter α for 0 ≤ β ≤ 900
α : continuum solution, α_s : Southwell, Δ : Southwell's error

β	α_s	α	Δ	β	α_s	α	Δ	β	α_s	α	Δ
0.0000	1.0000	1.0000	0	0.1	1.1	1.295	15	30	31	44.86	31
0.0005	1.0005	1.0015	0	0.2	1.2	1.580	24	40	41	57.54	29
0.001	1.001	1.0030	0	0.3	1.3	1.856	30	50	51	69.99	27
0.002	1.002	1.0060	0	0.4	1.4	2.123	34	60	61	82.27	26
0.003	1.003	1.0090	1	0.5	1.5	2.382	37	70	71	94.40	25
0.004	1.004	1.0120	1	0.6	1.6	2.633	39	80	81	106.44	24
0.005	1.005	1.0150	1	0.7	1.7	2.878	41	90	91	118.38	23
0.006	1.006	1.0180	1	0.8	1.8	3.116	42	100	101	130.25	22
0.007	1.007	1.0210	1	0.9	1.9	3.349	43	120	121	153.79	21
0.008	1.008	1.0240	2	1.0	2.0	3.576	44	140	141	177.13	20
0.009	1.009	1.0270	2	2.0	3.0	5.624	47	160	161	200.30	20
0.010	1.010	1.0300	2	3.0	4.0	7.427	46	180	181	223.33	19
0.020	1.020	1.0598	4	4.0	5.0	9.100	45	200	201	246.24	18
0.030	1.030	1.0896	5	5.0	6.0	10.697	44	300	301	359.51	16
0.040	1.040	1.1192	7	6.0	7.0	12.241	43	400	401	471.28	15
0.050	1.050	1.1487	9	7.0	8.0	13.749	42	500	501	581.79	14
0.060	1.060	1.1782	10	8.0	9.0	15.227	41	600	601	684.82	12
0.070	1.070	1.2075	11	9.0	10.0	16.683	40	700	701	765.61	8
0.080	1.080	1.2367	13	10.0	11.0	18.118	39	800	801	842.45	5
0.090	1.090	1.2659	14	20.0	21.0	31.820	34	900	901	912.66	1

For the sake of comparison, both the approximate values given by Eqn (7.28) and the exact values resulting from Eqn (7.24) are shown in Figs 7-7 and 7-8. This comparison also demonstrates that Southwell's additive theorem [formula (7.28)] is conservative. The maximum deviation from the exact value is 47%.

Formula (7.24) is not applicable in the special case when $EI_{\omega}=0$, i.e., when the substitute column only has torsional stiffness. The denominator of the fraction in Eqn (7.25) is zero and the value of the fraction goes to infinity. Equation (7.24) assumes the form 0x∞ and also collapses. In this particular case the critical load is obtained by making use of Southwell's theorem, i.e., from Eqn (7.28) as

$$N_{cr,\varphi} = \frac{1}{i_p^2} GJ\ . \tag{7.29}$$

It is worth having a close look at this result. It is often considered in structural engineering practice that the value of the critical load for a column will decrease with increasing height but Eqn (7.29) shows that the critical load is the same for a small and for a tall cantilever. In fact, its value does not even depend on the height of the column, neither on the distribution of the load along the bar. This fact must be borne in mind when designing columns with thin-walled closed cross-section. The practical case when an unstable system of columns is supported by a single core of closed cross-section of thin-walled elements falls into this category.

Of the three principal critical loads $N_{cr,x}$, $N_{cr,y}$ and $N_{cr,\varphi}$, the smallest one is of practical interest.

7.3.3 Singly-symmetric Layout

With monosymmetric cross-sections the shear centre lies on the axis of symmetry. Assume that axis x is an axis of symmetry and the centre of the vertical load lies on this axis. In this case $y_o=0$ holds and, after the reduction of the order of Eqs (7.9), (7.10) and (7.11), the governing differential equations assume the form

$$EI_y u''' + Nu' = 0, \tag{7.30}$$

$$EI_x v''' + N(v' - x_o \varphi') = 0, \tag{7.31}$$

$$EI_\omega \varphi''' - \left(GJ - N\frac{I_o}{A} \right) \varphi' - Nx_o v' = 0. \tag{7.32}$$

The boundary conditions for these equations come from Eqs (7.12)...(7.18) as

$$u(0) = v(0) = \varphi(0) = 0, \tag{7.33a}$$

$$u'(H) = v'(H) = \varphi'(H) = 0, \tag{7.33b}$$

$$u''(0) = v''(0) = \varphi''(0) = 0. \tag{7.33c}$$

The first equation does not contain the angle of rotation φ, which shows that buckling in the plane of symmetry develops without torsion. In this case the value of the critical load in the plane of symmetry is given by the formula

$$N_{cr,x} = qH = \frac{7.837EI_y}{H^2} . \tag{7.34}$$

Eqs (7.31) and (7.32) contain both v and φ, indicating that perpendicular to the plane of symmetry buckling occurs by a combination of bending of the axis of the bar in plane yz and twist of the cross-section about the shear centre axis. This phenomenon is torsional-flexural buckling. The value of the critical load for this torsional buckling results from the solution of the eigenvalue problem defined by Eqs (7.31) and (7.32).

The system of simultaneous equations has no solution of closed form and some approximate method is to be used. Approximate solutions are available at different levels. The Föppl-Papkovics theorem [Chapter 8] offers a simple approximation. According to the theorem, applied to this case, a conservative estimate of the critical load of torsional-flexural buckling is obtained by combining the critical load for flexural buckling in plane *yz* and that of pure torsional buckling:

$$N_{cr} = \left(\frac{1}{N_{cr,y}} + \frac{1}{N_{cr,\varphi}} \right)^{-1} . \tag{7.35}$$

In Eqn (7.35) $N_{cr,y}$ is the critical load for flexural buckling in plane *yz*, defined by formula (7.23), $N_{cr,\varphi}$ is the critical load for pure torsional buckling [formula (7.24)] and N_{cr} belongs to the (combined) torsional-flexural buckling. The approximation offered by formula (7.35) is good enough in many practical applications. However, in certain cases, when the building has a delicate support system or when more economic structure is to be achieved, better approximation is needed. As a rule, better results can be achieved by considering the original governing equations of the problem [Eqs (7.31) and (7.32)] and applying some mathematical procedure to them. Bessel's functions, Fourier series, the finite difference method and the finite element method are all obvious choices. However, earlier research showed that the resulting matrix of coefficients is often ill-conditioned and consequently the critical load is not reliable in certain cases when the structure has extreme stiffness characteristics. The problem lies with pure torsional buckling which, as mentioned earlier, is characterized with a differential equation identical with the governing equation of regular frameworks discussed in Chapter 5. Section 5.6 is devoted to the interesting role that the different stiffness characteristics play on the behaviour of the structure. Investigations into the stability of regular frameworks showed that the power series method, often considered outdated, could be successfully applied to structures with extreme stiffness characteristics. As the application of the generalized power series method is relatively simple and the method can be used for other structural engineering problems as well, the main points of the procedure will be presented in the following. Details of the method are available in [Zalka, 1981].

The task has been made easier as formula (7.35) offers a lower bound to the eigenvalue. Better convergence is ensured by introducing non-dimensional quantities:

$$\alpha = \frac{qH^3 i_p^2}{7.837EI_\omega} = \frac{N_{cr}}{N^{\omega}_{cr,\varphi}} , \tag{7.36}$$

$$\beta = \frac{GJH^2}{7.837EI_\omega} = \frac{N^t_{cr,\varphi}}{N^\omega_{cr,\varphi}}, \tag{7.37}$$

$$\gamma = \frac{EI_\omega}{EI_x i_p^2} = \frac{N^\omega_{cr,\varphi}}{N_{cr,y}}, \tag{7.38}$$

$$\tau = \frac{x_o}{i_p} = \sqrt{\frac{I_o - I_x - I_y}{I_o}}, \tag{7.39}$$

In the above formulae

$$N_{cr} = q_{cr}H$$

which is the critical load for combined torsional and flexural buckling, the eigenvalue of the problem,

$$i_p = \sqrt{\frac{I_o}{A}} = \sqrt{\frac{I_x + I_y + Ax_o^2}{A}} \tag{7.40}$$

which is the radius of gyration and $N_{cr,y}$, $N^t_{cr,\varphi}$ and $N^\omega_{cr,\varphi}$ are the part critical loads defined by formulae (7.23), (7.26) and (7.27). With the above notations, the differential equations of the problem assume the form:

$$v''' + \alpha\gamma \frac{7.837}{H^3} z(v' - i_p\tau\varphi') = 0, \tag{7.41}$$

$$\varphi''' - \left(\beta\frac{7.837}{H^2} - \alpha\frac{7.837}{H^3}z\right)\varphi' - \alpha\frac{7.837\tau}{i_p H^3}zv' = 0, \tag{7.42}$$

where α is the parameter of the eigenvalue. The boundary conditions are still given by Eqs (7.33a/b/c). No closed form solution is available for these differential equations but the power series

$$v = b_o + \sum_{k=1}^{\infty} b_k z^k, \tag{7.43}$$

and

$$\varphi = c_o + \sum_{k=1}^{\infty} c_k z^k \tag{7.44}$$

offer good approximations, which are easy to handle and which do not create difficulties later when convergency is of vital importance. Coefficients b_k and c_k in Eqs (7.43) and (7.44) are yet unknown. First, the necessary differential quotients of v and φ have to be substituted into the simultaneous differential equations (7.41) and (7.42), then it should be ensured that the equations equal zero. This is only possible if the sum of the coefficients of each power of z equals zero. Based on these conditions, the following formulae can be generated for the successive determination of the unknown coefficients:

$$b_k = -\frac{7.837\alpha\gamma(k-3)}{(k-2)(k-1)kH^3} b_{k-3} + \frac{7.837\alpha\gamma\tau i_p(k-3)}{(k-2)(k-1)kH^3} c_{k-3}, \tag{7.45}$$

$$c_k = \frac{7.837\beta}{(k-1)kH^2} c_{k-2} - \frac{7.837\alpha(k-3)}{(k-2)(k-1)kH^3} c_{k-3} + \frac{7.837\alpha\tau(k-3)}{(k-2)(k-1)k i_p H^3} b_{k-3}, \tag{7.46}$$

where $k \geq 3$. The initial values needed for the calculation of the coefficients are obtained from the boundary conditions (7.33a) and (7.33c), by making use of the power series (7.43) and (7.44) and their differential quotients. Thus:

$$b_o = c_o = 0, \tag{7.47}$$

$$b_2 = c_2 = 0. \tag{7.48}$$

The formulae for the successive determination of the unknown coefficients (7.45) and (7.46) show that coefficients b_k and c_k can be expressed as a function of b_1 and c_1, respectively:

$$b_k = A_{k,11} b_1 + i_p A_{k,12} c_1 , \qquad k \geq 3 \tag{7.49}$$

$$c_k = \frac{1}{i_2} A_{k,21} b_1 + A_{k,22} c_1 . \qquad k \geq 3 \tag{7.50}$$

In formulae (7.49) and (7.50), $A_{k,ij}$ are functions of α, β, γ, τ and k:

$$A_{k,ij} = f(\alpha,\beta,\gamma,\tau,k) . \qquad i,j = 1,2 \tag{7.51}$$

By making use of the last boundary condition (7.33b), the following system of "conditional" equations can be set up:

$$b_1 + \sum_{k=3}^{\infty} k b_k H^{k-1} = 0, \tag{7.52}$$

$$c_1 + \sum_{k=3}^{\infty} k c_k H^{k-1} = 0. \tag{7.53}$$

Equations (7.49) and (7.50) show that this system of "conditional" equations can be considered a linear combination of b_1 and c_1:

$$A_{11} b_1 + i_p A_{12} c_1 = 0 \tag{7.54}$$

$$\frac{1}{i_p} A_{21} b_1 + A_{22} c_1 = 0, \tag{7.55}$$

or in matrix form

$$\mathbf{A}\mathbf{x} = \mathbf{0}, \tag{7.56}$$

where

$$\mathbf{A} = \begin{bmatrix} A_{11} & i_p A_{12} \\ \frac{1}{i_p} A_{21} & A_{22} \end{bmatrix} \quad \text{and} \quad \mathbf{x} = \begin{bmatrix} b_1 \\ c_1 \end{bmatrix}. \tag{7.57}$$

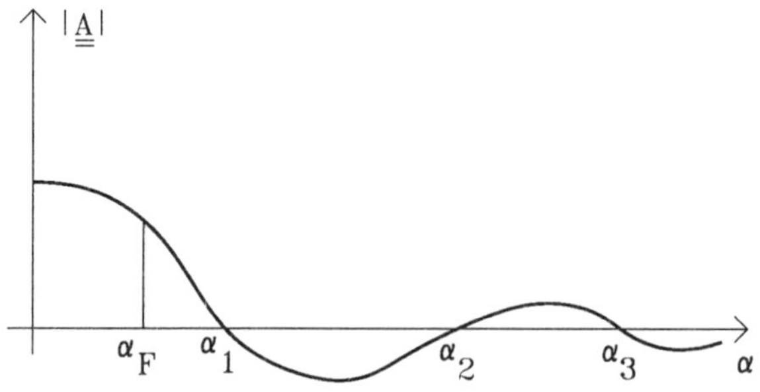

Fig. 7-9. Polynomial of *n*-th degree for the eigenvalue problem

Coefficients b_1 and c_1 are only equal to zero in the trivial case and in every other case they are different from zero. It follows that from the homogeneous, linear equation (7.56), the expression

$$| \mathbf{A} | = 0 \tag{7.58}$$

yields the non-trivial solution.

Every element of matrix A results from an infinite summation. According to an accuracy analysis covering structures with extreme stiffness characteristics and with known results available from other sources, it is enough to take into consideration the first n terms. Setting an error level of 10^{-12}, the number of necessary terms, i.e., the value of n is automatically determined. Matrix A only contains α as an unknown quantity and when the value of its determinant is set equal to zero, a polynomial of the n-th degree is obtained in α.

The roots of this polynomial are the eigenvalues of the eigenvalue problem. Only the smallest root is needed because only the smallest eigenvalue, i.e., the critical load is of practical importance. The determination of the critical load is made easier as the shape of the polynomial is known: Due to the physical nature of the problem - only structures with non-negative geometrical and stiffness characteristics are dealt with -, the polynomial assumes a positive value at $\alpha=0$ (Fig. 7-9).

The eigenvalue problem of the system of differential equations of variable coefficients for combined flexural and torsional buckling has thus been simplified: The value of the critical load is determined as the smallest root of a polynomial of n-th degree.

Table 7-2. Critical load parameter ε for monosymmetric cross-sections

$\dfrac{N_{cr,\varphi}}{N_{cr,y}}$	0.5	1.0	2.0	3.0	4.0	5.0
τ = 0.0	0.500	1.000	1.000	1.000	1.000	1.000
τ = 0.1	0.496	0.910	0.986	0.992	0.998	1.000
τ = 0.2	0.483	0.834	0.962	0.980	0.992	0.993
τ = 0.3	0.464	0.770	0.926	0.956	0.977	0.981
τ = 0.4	0.443	0.715	0.882	0.925	0.962	0.966
τ = 0.5	0.424	0.667	0.841	0.895	0.936	0.947
τ = 0.6	0.403	0.626	0.800	0.862	0.913	0.927
τ = 0.7	0.384	0.589	0.765	0.832	0.885	0.905
τ = 0.8	0.365	0.556	0.728	0.802	0.853	0.881
τ = 0.9	0.349	0.527	0.699	0.775	0.826	0.858
τ = 1.0	0.333	0.500	0.667	0.750	0.800	0.833

The necessary calculations were carried out and the Föppl-Papkovics approximate solution (7.35) was used as a lower bound for the successive approximation (α_F in Fig. 7-9). Equation (7.58) was solved for different geometrical and stiffness characteristics. With an error level 10^{-12} set for the calculations, the results presented in Fig. 7-10 and in Table 7-2 can be considered exact for practical structural engineering applications.

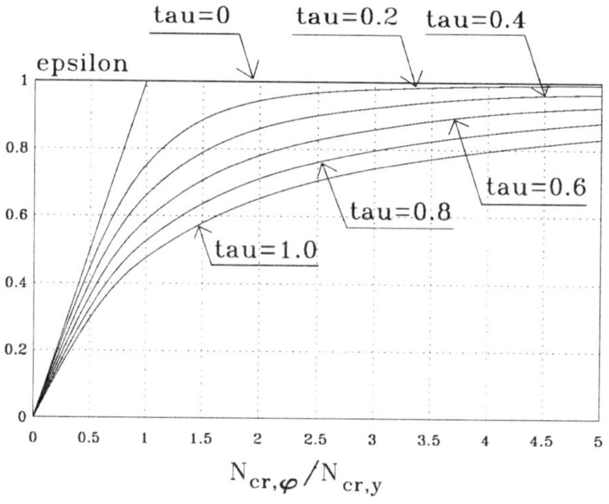

Fig. 7-10. Critical load parameter ε for monosymmetric cross-sections

The critical load for combined torsional-flexural buckling of structures with monosymmetric cross-section can be computed as follows. First, the critical load for buckling in the plane perpendicular to the plane of symmetry is calculated:

$$N_{cr,y} = qH = \frac{7.837EI_x}{H^2} . \tag{7.59}$$

This flexural buckling combines with pure torsional buckling. The critical load for pure torsional buckling $N_{cr,\varphi}$ is calculated from Eqn (7.24). Details of this procedure are presented in Section 7.3.2. Knowing $N_{cr,y}$ and $N_{cr,\varphi}$ and taking into consideration stiffness ratio τ defined by formula (7.39), the combined critical load is calculated from the formula

$$N_{cr} = \varepsilon N_{cr,y} , \tag{7.60}$$

where values for parameter ε are given in Table 7-2 and Fig. 7-10.

Buckling in the plane of symmetry does not combine with pure torsional buckling. This critical load $N_{cr,x}$ is produced by formula (7.34). The actual critical load - the one which is then compared to the applied loads during the design procedure - is of course the smaller one of N_{cr} and $N_{cr,x}$.

It is worth analysing the two extreme values of parameter τ in detail.

Value τ=0 corresponds to the case when x_o=0, i.e., when the cross-section is doubly symmetric. Failure can occur by pure torsional *or* flexural buckling. This fact is also shown by the two intersecting lines in Fig. 7-10. While the critical load for pure torsional buckling is smaller than the critical load for flexural buckling, the actual critical load equals the former one. On the other hand, when the critical load for pure torsional buckling is greater than the critical load for flexural buckling, the actual critical load equals the latter one.

Value τ=1.0 correspond to the case when the shear centre is at infinity. It is interesting to note that the values represented by the lowest curve in Fig. 7-10 are identical with those given by the Föppl-Papkovics approximate formula, in other words, in this case the approximate formula results in the exact values.

7.3.4 Unsymmetric Layout

When the cross-section has no axis of symmetry or it has (one or two) symmetry axes but the centroid of the applied vertical loads does not lie on the symmetry axes, pure torsional buckling combines with flexural buckling in both principal planes.

First, an approximate formula will be presented for the critical load of the torsional buckling of a cantilever with unsymmetric cross-section. The Föppl-Papkovics additive theorem results in the following simple formula:

$$N_{cr} = \left(\frac{1}{N_{cr,x}} + \frac{1}{N_{cr,y}} + \frac{1}{N_{cr,\varphi}} \right)^{-1} . \tag{7.61}$$

This formula is the three dimensional version of formula (7.35). As is the case with formula (7.35), formula (7.61) is also conservative and is applicable to many structural engineering problems. In certain cases, however, when the three part critical loads $N_{cr,x}$, $N_{cr,y}$ and $N_{cr,\varphi}$ are of the same order of magnitude and parameters τ_x and τ_y in formulae (7.66) and (7.67) are small, it may result in a critical load with too great an approximation on the safe side. In such cases, the exact solution should be used. This procedure will now be presented.

The combined flexural and torsional buckling of a cantilever with unsymmetric cross-section, subjected to uniformly distributed normal load is characterized by the differential equations (7.9), (7.10) and (7.11) and the boundary conditions (7.12),...,(7.18).

Introduce the following non-dimensional quantities:

$$\alpha = \frac{qH^3 i_p^2}{7.837EI_\omega} = \frac{N_{cr}}{N^{\omega}_{cr,\varphi}} , \tag{7.62}$$

$$\beta = \frac{GJH^2}{7.837EI_\omega} = \frac{N^t_{cr,\varphi}}{N^\omega_{cr,\varphi}}, \tag{7.63}$$

$$\delta = \frac{EI_\omega}{EI_y i_p^2} = \frac{N^\omega_{cr,\varphi}}{N_{cr,x}}, \tag{7.64}$$

$$\gamma = \frac{EI_\omega}{EI_x i_p^2} = \frac{N^\omega_{cr,\varphi}}{N_{cr,y}}, \tag{7.65}$$

$$\tau_x = \frac{x_o}{i_p} = \sqrt{\frac{I_o - I_x - I_y - y_o^2 A}{I_o}}, \tag{7.66}$$

$$\tau_y = \frac{y_o}{i_p} = \sqrt{\frac{I_o - I_x - I_y - x_o^2 A}{I_o}}, \tag{7.67}$$

where

$$i_p^2 = \frac{I_o}{A} = \frac{I_x + I_y + x_o^2 A + y_o^2 A}{A}. \tag{7.68}$$

With these notations and after a reduction of order, the governing differential equations (7.9), (7.10), (7.11), assume the following form:

$$u''' + \alpha\delta \frac{7.837}{H^3} z(u' + i_p \tau_y \varphi') = 0, \tag{7.69}$$

$$v''' + \alpha\gamma \frac{7.837}{H^3} z(v' - i_p \tau_x \varphi') = 0, \tag{7.70}$$

$$\varphi''' - \left(\beta\frac{7.837}{H^2} - \alpha\frac{7.837}{H^3}z\right)\varphi' - \alpha z\frac{7.837}{H^3 i_p}\left(\tau_x v' - \tau_y u'\right) = 0. \tag{7.71}$$

Expressions (7.33a/b/c) hold the boundary conditions. The same method shown in detail in Section 7.3.3 where monosymmetric cross-sections were dealt with can be applied. The solution can be found again by using infinite power series, this time three series, covering deflection in both principal planes and rotation around the shear centre axis. First, the three series and their differential quotients should be substituted into the governing differential equations. This leads to three equations for the successive determination of the unknown coefficients of the power series. Two of the three sets of boundary conditions result in the necessary initial values for the recursion and the third set makes it possible to establish a set of three linear algebraic equations. These are the "conditional" equations, which are analogous with Eqs (7.54) and (7.55). Apart from the remaining three unknown coefficients, the equations only contain the critical load parameter, the eigenvalue of the problem. In setting the determinant of the equations equal to zero, one obtains the critical load parameter as the smallest root of the resulting polynomial.

The outlined procedure was carried out with different stiffness ratios. The load parameters needed for the calculation of the critical load are given in Tables 8-1,...,8-10 in Chapter 8.

Using the Tables, the critical load can be calculated as follows.

The value of τ_x [Eqn (7.66)] shows which Table is to be used. As a function of the ratios

$$\frac{N_{cr,\varphi}}{N_{cr,x}} \quad \text{and} \quad \frac{N_{cr,\varphi}}{N_{cr,y}} \tag{7.72}$$

and the ratio τ_y [Eqn (7.67)], the load parameter

$$\lambda = \frac{N_{cr}}{N_{cr,\varphi}} \tag{7.73}$$

can be found in one of the Tables. The value of the critical load for pure torsional buckling $N_{cr,\varphi}$ is obtained by making use of Eqn (7.24) and Table 7-1 or Figs 7-7 and 7-8. The other two part critical loads $N_{cr,x}$ and $N_{cr,y}$ are defined by Eqs (7.22) and (7.23). Knowing the load parameter λ, the following formula gives the value of the critical load:

$$N_{cr} = q_{cr}H = \lambda N_{cr,\varphi} \, . \tag{7.74}$$

To make the calculation of the load parameter reasonably simple, the values of τ_x are presented in the Tables in increments of 0.1 and the values of τ_y in increments of 0.2. In this way there are ten tables to be used. The values of τ_y in each table are placed in successive lines so that interpolation is simple. The Tables contain a large number of values for the load parameter so that linear interpolation is considered accurate enough for practical application.

The following extreme cases are worth mentioning.

The case $\tau_x=\tau_y=0$ corresponds to structures with doubly symmetric cross-sections. This case was analysed in Section 7.3.2 in detail.

When $\tau_y=0$ and $\tau_x\neq0$ or $\tau_x=0$ and $\tau_y\neq0$, the structure is monosymmetric.

In these cases flexural buckling in the plane of symmetry is independent from the combined torsional-flexural buckling in the plane perpendicular to the axis of symmetry. The load parameters in the Tables take this fact into consideration and they always result in the actual critical load. For example, with fixed $\tau_x=0.9$, if

$$\frac{N_{cr,\varphi}}{N_{cr,y}} = 0.8,$$

then 0.585 is obtained from Table 8-9 for the value of

$$\lambda = \frac{N_{cr}}{N_{cr,\varphi}},$$

until

$$\frac{N_{cr,\varphi}}{N_{cr,x}} \leq 1$$

holds, since flexural bending in plane *xy* is independent of torsional flexural buckling in plane *yz*. After exceeding

$$\frac{N_{cr,\varphi}}{N_{cr,x}} = 1,$$

the values of the critical load parameter for buckling in the plane of symmetry become lower than those of the torsional-flexural ones, so that the tables give the lower values.

When $\tau_x=\tau_y=0$, both coordinates of the shear centre are at infinity. This is the case when the Föppl-Papkovics additive formula (7.61) results in the best approximation. It is interesting to mention, however, that there are no geometrical and stiffness ratios in this unsymmetric case when the additive formula results in the exact critical load - contrary to the monosymmetric case, when formula (7.35) provides the exact solution for $\tau=1$.

The values of the Tables show that for structures with identical $N_{cr,x}$, $N_{cr,y}$, $N_{cr,\varphi}$ the bigger τ_x and τ_y are, the smaller the combined torsional-flexural load is. Even a small change in the value of τ_x and τ_y can cause considerable change in the value of the critical load indicating that the distance between the shear centre and the gravity centre plays an important role in the stability of the structure.

The values in the Tables show certain symmetry properties. If, for example, τ_x equals τ_y, the corresponding load factors are symmetric to the diagonal of the Table. This follows from the fact that flexural buckling in planes *xz* and *yz* is the same physical phenomenon. The governing differential equations (7.9), (7.10) also show this fact.

The method presented above makes it possible to save considerable time and energy in carrying out the stability analysis of even large structures. The task can be further reduced by developing a computer procedure, which solves the governing differential equations and produces the critical load for the actual structure directly, without making it necessary to use the Tables. Such a computer program has been developed [Fazakas et al, 1989] which, after receiving the geometrical characteristics, also carries out the often cumbersome job of computing the individual and overall stiffness characteristics. By making use of the program, the stability analysis of even large structures can be carried out in minutes.

7.4 CONCENTRATED LOAD AT ROOF LEVEL

There are practical cases when, in addition to the UDL on the floors, there is some concentrated force at top floor level. This is the case when there is a panorama restaurant, a swimming pool at top floor level or only the dead weight of the top floor structure is greater than that of the intermediate floor structures.

When the substitute column is only subjected to a concentrated force, the eigenvalue problem of governing differential equations assumes a simple form [Timoshenko and Gere, 1961]:

$$EI_y u'''' + F\left(u' + y_o \varphi'\right)' = 0, \tag{7.75}$$

$$EI_x v'''' + F\left(v' - x_o \varphi'\right)' = 0, \tag{7.76}$$

$$EI_\omega \varphi'''' - \left(GJ\varphi' - F\frac{I_o}{A}\varphi'\right)' - F\left(x_o v' - y_o u'\right)' = 0. \tag{7.77}$$

The critical load is given as the smallest root of a cubic polynomial originating from the determinant

$$\begin{vmatrix} (F_{cr} - F_{cr,x}) & 0 & F_{cr} y_o \\ 0 & (F_{cr} - F_{cr,y}) & -F_{cr} x_o \\ F_{cr} y_o & -F_{cr} x_o & i_p^2 (F_{cr} - F_{cr,\varphi}) \end{vmatrix} \tag{7.78}$$

The following notations are used in determinant (7.78):

$$F_{cr,x} = \frac{\pi^2 EI_y}{4H^2} \tag{7.79}$$

is Euler's critical load in plane xz,

$$F_{cr,y} = \frac{\pi^2 EI_x}{4H^2} \tag{7.80}$$

is Euler's critical load in plane *yz* and

$$F_{cr,\varphi} = \frac{1}{i_p^2}\left(GJ + \frac{\pi^2 EI_\omega}{4H^2}\right) \tag{7.81}$$

is the critical load for pure torsional buckling.

The geometrical and stiffness characteristics in formulae (7.79), (7.80) and (7.81) are defined in Section 7.2.

7.5 PRACTICAL APPLICATION

Formula (7.74) is the exact solution for a cantilever which is subjected to uniformly distributed vertical load along its entire length. It can be directly used for the torsional-flexural buckling analysis of a thin-walled cantilever subjected to its own weight.

With buildings, however, the actual load is somewhat different and certain considerations are needed.

7.5.1 Formulae for the Overall Analysis of Buildings

With multistorey buildings, the applied vertical load consists of concentrated forces at floor levels, each representing the load of one storey. The UDL of the substitute column is obtained by distributing these forces downwards along the column (Fig. 7-5). In this way, a model, representing a more favourable situation than the original one, is created. This step leads to an overestimated critical load. The same phenomenon emerged in Chapter 5 when the substitute column of frameworks on fixed supports was established. As it was explained in detail in Section 5.5.2, a possible correction can be made by the application of Dunkerley's additive theorem. According to the additive formula, the critical load is obtained from

$$N_{cr}^D = \left(\frac{1}{2eF_{cr}} + \frac{1}{N_{cr}}\right)^{-1}. \tag{7.82}$$

In formula (7.82), F_{cr} is the critical load for the concentrated load on top (solution of 7.78), N_{cr} is the critical load for UDL (Eqn 7.74), and e is the number of storeys.

It is also possible to take into consideration both the concentrated forces at top floor level and UDL on the floors. Knowing the critical load for concentrated forces at top level and UDL on the floor levels, the formula

$$\nu = \frac{1}{\dfrac{F}{F_{cr}} + \dfrac{N}{N_{cr}}} . \tag{7.83}$$

gives the ratio of the critical load to the applied load. In formula (7.83) F and N are the applied concentrated load at top floor level and the sum of the UDL at floor levels.

7.5.2 Design Examples

Two typical application possibilities are presented in this section. First, through the analysis of a multistorey building supported by several cores, it will be shown in detail how the actual computation can be carried out by using the formulae and Tables given in this Chapter.

Second, using Kollår's [1977] excellent example, it will be demonstrated how sensitive a building can be to the arrangement of the support system - in this case a single core.

7.5.2.1 System of Support Cores

Using the formulae derived for the critical load, the stability of the fourteen-storey building shown in Fig. 7-11 is now investigated.

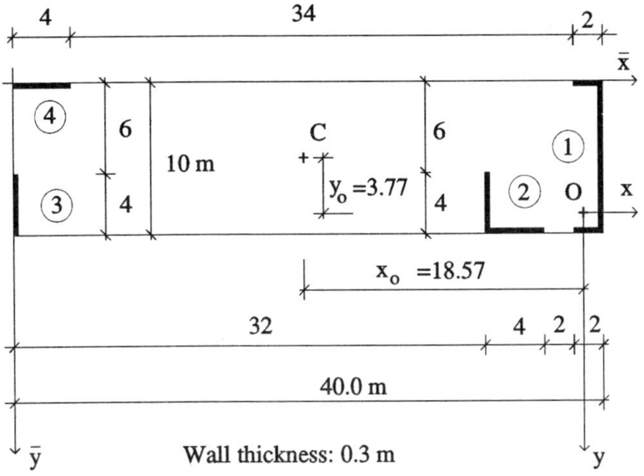

Fig. 7-11. Ground plan for numerical example 7.5.2.1

The building used for this numerical analysis is identical to the one used by Beck to check the accuracy of different methods developed for the stress analysis of buildings [Beck and Shäfer, 1969]. The load bearing system of the building consists of pin-ended columns, U and L shaped cores and shear walls. The pin-ended columns are not taken into account for the analysis.

The characteristic data for the analysis are as follows.

Size of the ground plan: L = 40.0 m, B = 10.0 m .
Storey height: h = 2.86 m. Number of storeys: e = 14 .
Height of the building: H = 40.0 m.
Modulus of elasticity: E = 25,000 MN/m^2.
Shear modulus: G = 12,500 MN/m^2.
Concentrated load at the top of the building: F = 1.0 MN .
UDL on the floors: q = 10.0 kN/m^2.
Total vertical load on the floors: N = 14×40×10×0.010 = 56.0 MN .

The stiffness characteristics are collected in Table 7-3.

First, the location of shear centre O is determined from formulae (7.1) and (7.2):

$$\bar{x}_o = 38.565 \text{ m}$$

and

$$\bar{y}_o = 8.771 \text{ m} .$$

Note that the above coordinates of the shear centre are in the co-ordinate system whose axes border the ground plan (Fig. 7-1). After transferring the origin of the coordinate system to the shear centre of the support system, the distance between the centroid of the ground plan (C) and the shear centre is computed:

$$t = \sqrt{\left(\bar{x}_o - \frac{L}{2}\right)^2 + \left(\bar{y}_o - \frac{B}{2}\right)^2} = 18.944 \text{ m} .$$

Table 7-3. Stiffness characteristics for numerical example 7.5.2.1

Core	$\bar{x}_i$	$\bar{y}_i$	A_i	$I_{x,i}$	$I_{y,i}$	$I_{xy,i}$	J_i	W_i	W
1	40.34	5.00	4.02	48.93	1.00	0.00	0.121	17.87	
2	32.15	9.85	2.31	3.57	3.58	2.14	0.069	0.17	
3	0.15	8.00	1.20	1.60	0.01	0.00	0.036	0.00	
4	2.00	0.15	1.20	0.01	1.60	0.00	0.036	0.00	
Σ			8.73	54.11	6.19	2.14	0.262		2860.1

The critical load for concentrated forces on top of the building will now be computed.

The critical loads for flexural buckling in planes xz and yz are given by Euler's formulae (7.79) and (7.80), respectively:

$$F_{cr,x} = \frac{\pi^2 EI_y}{4H^2} = \frac{\pi^2 \times 25{,}000 \times 6.19}{4 \times 40^2} = 238.6 \text{ MN}$$

and

$$F_{cr,y} = \frac{\pi^2 EI_x}{4H^2} = \frac{\pi^2 \times 25{,}000 \times 54.11}{4 \times 40^2} = 2086.2 \text{ MN} .$$

For the calculation of the critical load for pure torsional buckling, the value of the radius of gyration (Eqn 7-8) is needed:

$$i_p = \sqrt{\frac{(L^2 + B^2)}{12} + t^2} = \sqrt{\frac{40^2 + 10^2}{12} + 18.944^2} = 22.37 \text{ m} .$$

The critical load for pure torsional buckling is obtained from Eqn (7.81):

$$F_{cr,\varphi} = \frac{1}{i_p^2}\left(GJ + \frac{\pi^2 EI_\omega}{4H^2}\right) = \frac{12{,}500 \times 0.262}{500.5} + \frac{\pi^2 \times 25{,}000 \times 2860.1}{500.5 \times 4 \times 40^2} =$$

$$= 6.5 + 220.3 = 226.8 \text{ MN} .$$

With $F_{cr,x}$, $F_{cr,y}$ and $F_{cr,\varphi}$, the critical load is given as the smallest root of a cubic polynomial originating from determinant (7.78):

$$F_{cr} = 191.4 \text{ MN} .$$

The calculation of the critical load for UDL on the floors is now presented.

The critical load for buckling in plane xz (Eqn 7.22) assumes the value:

$$N_{cr,x} = \frac{7.837 EI_y}{H^2} = \frac{7.837 \times 25{,}000 \times 6.19}{40^2} = 758.0 \text{ MN} .$$

The critical load for buckling in plane yz (Eqn 7.23) is

$$N_{cr,y} = \frac{7.837 EI_x}{H^2} = \frac{7.837 \times 25{,}000 \times 54.1}{40^2} = 6624.7 \text{ MN} .$$

For the calculation of the critical load for pure torsional buckling, the part critical load associated with the torsional stiffness (Eqn 7.26) and the part critical load associated with the warping stiffness (Eqn 7.27) are needed. They assume the values:

$$N^{t}_{cr,\omega} = \frac{GJ}{i_p^2} = \frac{12{,}500\times 0.262}{22.37^2} = 6.5 \text{ MN}$$

and

$$N^{\omega}_{cr,\varphi} = \frac{7.837\,EI_\omega}{i_p^2 H^2} = \frac{7.837\times 25{,}000\times 2{,}860.1}{22.37^2\times 40^2} = 699.9 \text{ MN} .$$

The ratio of the above part critical loads (Eqn 7.25) is

$$\beta = \frac{N^{t}_{cr,\varphi}}{N^{\omega}_{cr,\varphi}} = \frac{6.5}{699.9} = 0.0093 .$$

The critical load parameter for pure torsional buckling is obtained from Table 7-1 in the function of β:

$$\alpha = 1.028 .$$

With the load parameter and the part critical load for the warping stiffness, formula (7.24) yields the critical load for pure torsional buckling:

$$N_{cr,\varphi} = \alpha N^{\omega}_{cr,\varphi} = 1.028\times 699.9 = 719.5 \text{ MN} .$$

The centroid of the vertical load does not coincide with the shear centre and it is not on either principal axes of inertia so that pure torsional buckling combines with flexural buckling in both principal planes. First, the values of ratios (7.66), (7.67) and (7.72) are needed

$$\tau_x = \frac{x_o}{i_p} = \frac{18.565}{22.37} = 0.830 ,$$

$$\tau_y = \frac{y_o}{i_p} = \frac{3.771}{22.37} = 0.169 ,$$

$$\frac{N_{cr,\varphi}}{N_{cr,x}} = \frac{719.5}{758.0} = 0.949 ,$$

$$\frac{N_{cr,\varphi}}{N_{cr,y}} = \frac{719.5}{6624.7} = 0.109 .$$

Then, after linear interpolation, Tables 8-8 and 8-9 in Chapter 8 give the load parameter as:

$$\lambda = 0.895 \ .$$

With the load parameter, the critical load of the substitute column subjected to UDL along the entire length is obtained from Eqn (7.74):

$$N_{cr} = \lambda N_{cr,\varphi} = 0.895\times 719.5 = 643.9 \text{ MN} \ .$$

By the application of Dunkerley's additive formula, the effect of the approximation following from the distribution of the load on the floors along the height of the building can be reduced. Formula (7.82) results in

$$N^{D}_{cr} = \left(\frac{1}{2eF_{cr}} + \frac{1}{N_{cr}}\right)^{-1} = \left(\frac{1}{2\times 14\times 191.4} + \frac{1}{643.9}\right)^{-1} = 574.9 \text{ MN}$$

for the critical load of the building subjected to UDL on the floors.

Finally, formula (7.83) gives the ratio of the overall critical load to the applied loads:

$$\nu = \frac{1}{\dfrac{F}{F_{cr}} + \dfrac{N}{N_{cr}}} = \frac{1}{\dfrac{1.0}{191.4} + \dfrac{56.0}{574.9}} = 9.7 \ .$$

According to this result, the building can theoretically withstand 9.7 times as much load as the currently applied load. For the actual structural design, the application of safety factors in accordance with the relevant code of practice may have to be considered.

7.5.2.2 A Single Support Core

Consider the five storey building with four different core arrangements shown in Figs 7-12 and 7-13 and investigate how the stability of the structure is affected when the support core is placed at different locations. The floors are supported by pinned columns. Again, the pinned columns are not taken into account in the stability analysis. The horizontal stiffness is provided by a single core. The basic geometrical, stiffness and loading characteristics common in all four cases are as follows.

UDL on the floors: $q = 8.0$ kN/m^2.
Size of the ground plan: $L = 26.0$ m, $B = 14.0$ m .
Storey height: $h = 3.0$ m. Number of storeys: $e = 5$.
Height of the building: $H = 15.0$ m.
Size of the core: $h = 3.2$ m, $b = 2.6$ m, $t = 0.15$ m.
Shear modulus: $G = 9{,}580$ MN/m^2.
Modulus of elasticity: $E = 23{,}000$ MN/m^2.
Concentrated load at the top of the building: $F = 1.0$ MN .
Total vertical load on the floors: $N = 26\times 14\times 5\times 0.008 = 14.56$ MN .

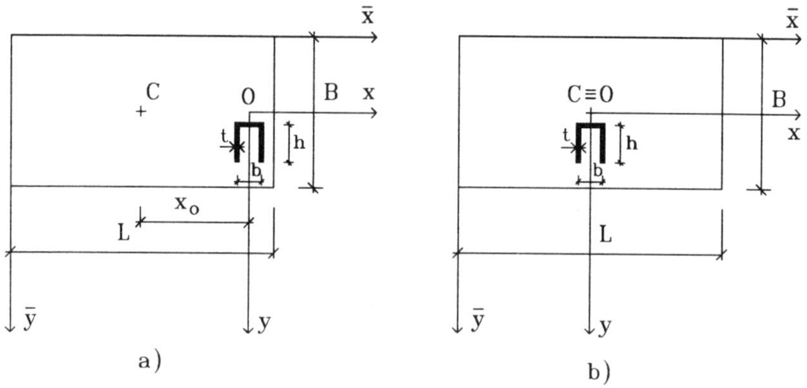

Fig. 7-12. Layouts for numerical example: a) case "A", b) case "B"

The geometrical and stiffness characteristics representing the four cases are given in Table 7.4. The part critical loads which contribute to the overall critical load are presented in Table 7-5 for the concentrated top load and in Table 7-6 for the UDL on the floors. All the critical loads are given in meganewton. Details of the computation carried out by using the computer procedure [Fazakas et al, 1989] are not presented in this numerical example.

Table 7-4. Geometrical and stiffness characteristics for numerical example 7.5.2.2

Case	I_x	I_y	I_{xy}	J	I_ω	x_o	y_o	i_p
"A"	1.53	1.84	0.00	0.01	1.880	11.00	0.00	13.92
"B"	1.53	1.84	0.00	0.01	1.880	0.00	0.00	8.52
"C"	2.15	1.92	0.00	2.93	0.023	11.00	3.00	14.24
"D"	2.15	1.92	0.00	2.93	0.023	0.00	0.00	8.52

Case "A"

First, the support core is located in the corner of the building (Fig. 7-12/a). The support core is a U-shaped core with the wall thickness of 0.15 m. The shear centre of the core lies on the centroidal axis *x*. Consequently, pure torsional buckling and flexural buckling in plane *yz* combine. The overall critical load assumes the value

$$F_{cr} = 2.93 \text{ MN}$$

when the building is only subjected to concentrated load at top floor level and

$$N_{cr} = 7.35 \text{ MN}$$

when the only load is uniformly distributed load on the floors. The ratio of the critical load to the applied loads is

$$\nu = 0.43 \ .$$

This ratio shows that the building is not stable.

Case "B"

Move the core to the centroid of the ground plan, i. e., to the centroid of the applied vertical loads (Fig. 7-12/b). The wall thickness is 0.15 m and the stiffness characteristics of the support system do not change. The only change is in the radius of gyration, whose value decreases (Table 7-4). The centre of the vertical loads and that of the support system coincide and therefore flexural buckling in the principal planes and pure torsional buckling are independent. The lowest value determines the overall critical load. According to the part critical loads in the second row in Table 7-5 and Table 7-6, the overall critical load is

$$F_{cr} = 7.86 \text{ MN}$$

when the building is only subjected to concentrated load at top floor level and

$$N_{cr} = 19.67 \text{ MN}$$

when the only load is uniformly distributed load on the floors. The ratio of the critical load to the applied loads is

$$\nu = 1.15 \ .$$

Table 7-5. Part critical loads for the concentrated load

Case	$F_{cr,x}$	$F_{cr,y}$	$F_{cr,\varphi}$	$F^{t}_{cr,\varphi}$	$F^{\omega}_{cr,\varphi}$
"A"	464.6	385.7	2.95	0.50	2.45
"B"	464.6	385.7	7.86	1.33	6.52
"C"	483.5	542.5	138.29	138.26	0.03
"D"	483.5	542.5	385.68	385.60	0.08

In theory, this value indicates a stable structure. However, according to the widely accepted structural engineering concept, a ratio of 4 is required for a structure which is considered stable enough.

Table 7-6. Part critical loads for UDL on the floors

Case	$N_{cr,x}$	$N_{cr,y}$	$N_{cr,\varphi}$	$N^{t}_{cr,\varphi}$	$N^{\omega}_{cr,\varphi}$
"A"	1120.0	929.7	7.38	0.50	5.90
"B"	1120.0	929.7	19.67	1.33	15.73
"C"	1165.6	1307.7	138.33	138.26	0.07
"D"	1165.6	1307.7	385.80	385.60	0.19

Case "C"

The location of the core is the same but the open section is now closed (Fig. 7-13/a). To allow for the window and door openings in the opposite walls parallel to axis x, a reduced wall thickness is introduced for the walls of length 2.6 m. An approximate computation based on a simple bending analysis results in an equivalent wall thickness of t'=0.1 m.

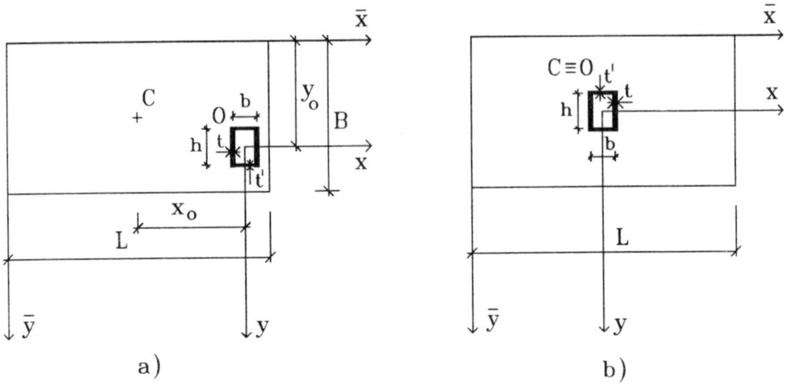

Fig. 7-13. Layouts for numerical example: a) case "C", b) case "D"

The bending stiffnesses change slightly, the warping stiffness nearly vanishes and the torsional stiffness increases considerably. The centre of the support system does not lie on either principal axes and so flexural buckling in both principal planes and pure torsional buckling combine. The resulting overall critical load is

$$F_{cr} = 117.3 \text{ MN}$$

when the building is only subjected to concentrated load at top floor level and

$$N_{cr} = 129.2 \text{ MN}$$

when the only load is uniformly distributed load on the floors. The ratio of the critical load to the applied loads now is

$$\nu = 8.25,$$

indicating a stable enough structure.

Case "D"

For the sake of comparison, place this closed core to the centroid of the vertical loads (Fig. 7-13/b). This move results in double symmetry when the three characteristic buckling modes can only develop independently and the smallest one is the resulting overall critical load. The smallest part critical load belongs to pure torsional buckling. According to Tables 7-5 and 7-6, the overall critical load assumes the value

$$F_{cr} = 385.7 \text{ MN}$$

when the building is only subjected to concentrated load at top floor level and

$$N_{cr} = 385.8 \text{ MN}$$

when the only load is uniformly distributed load on the floors. The ratio of the critical load to the applied loads is

$$\nu = 24.8 \ .$$

It is interesting to note that the overall critical load is practically identical in both load cases and its value does not depend on the nature of the load. Another fact, not shown in the Tables, is that the value of the critical loads in both load cases are independent of the height of the structure. This is always the case when buckling is ruled by pure torsional buckling and pure torsional buckling is ruled by the torsional stiffness while the warping stiffness is negligible (last row in Table 7-4).

7.6 SUPPLEMENTARY REMARKS

Regular structures have been investigated when the location of the shear centre is considered identical on every floor. It has also been assumed that the support elements are solid walls or cores without openings. These assumptions are widely accepted and used for approximate analysis in practical applications. However, further research in these areas would result in even better approximations and would widen the range of possible applications.

7.6.1 Shear Centre - Centre of Torsion

In introducing the substitute column in Section 7.1 it was assumed that the shear centre only depends on the geometrical characteristics of the cross-section. This - often tacit - assumption [Bornscheuer, 1952; Szmodits, 1975] is correct for open cross-sections. For closed cross-sections, however, this assumption does not hold. This follows from the fact that the above assumption is equivalent with the neglect of the shear deformation resulting from bending. Torsion does not result in this type of deformation in the case of open cross-sections. With closed cross-sections, however, the torsional deformation is of this kind and the shear centre and the centre of torsion do not coincide. According to investigations, the error resulting from the above approximation rarely exceeds 5 per cent [Pearson, 1956]. This error is even less if the support system of the building has cores with both open and closed cross-sections. No error is made, of course, when there are only cores with open cross-sections.

In taking into account shear deformations as well, the location of the shear centre also depends on the load and may vary along the axis of the bar [Stüssi, 1965]. Depending on the load distribution along the height of the structure, the location of the shear centre can vary considerably on the floor levels. This phenomenon may play an important role in the dynamic analysis of tall buildings when the load also varies in time [Kókai, 1983].

The shear centre also varies along the height of the structure when the arrangement of the cores on the floor levels are different. Some of the cores may not continue to the top of the structure as they are not needed for the support system or some of the cores are left out on ground floor level as the architect needs more space there. The differential equations of the problem have to be solved for all the different arrangements (horizontal problem) and then another procedure is needed to build the different levels together (vertical problem). Such structures cannot be considered regular and their stability analysis is beyond the scope of this book.

7.6.2 Pierced Support Elements

The support system of buildings often consists of pierced elements: Coupled shear walls and cores with openings. It is difficult to take into account the weakening effect of the openings. The state of stress of the structure becomes complex and the problem can only be defined by a large system of differential equations. This is why approximate methods of different accuracy are used in structural engineering practice.

The simplest and less accurate method is obtained when the effect of the beams linking the walls is neglected. The application of this method is justified when the structure is subjected to great loads (e.g. earthquake) which would make the beams fail anyway. Great advantage of this method is simplicity and the fact that it is conservative, i.e., it results in a lower bound to the critical load. The state of stress of the structure is clear and the effect of the individual stiffness characteristics is separated which makes it possible to strengthen the structure where it is most needed.

Since bars with thin-walled open cross-sections mainly resist torque by making use of their warping stiffness - and this phenomenon can be traced back to the bending of the elements of the cross-section [Vlasov, 1961; Zalka, 1980b] -, as an approximation, pierced shear walls and cores can be replaced by solid shear walls and cores with substitute thicknesses. The calculation of the substitute thickness can be based on the requirement of identical maximum deflection or identical average deflection. Well established methods are available for this calculation [Rosman, 1965; Irwin, 1984; Taranath, 1988]. In this way the effect of the axial deformation of the walls - considerable for tall buildings - can also be taken into account [Danay, Gellert and Glück, 1975]. Another possible approach may be the requirement of identical critical load.

A simple approximate method is available for the spatial stability analysis of a system of coupled shear walls, subjected to concentrated forces at top floor level. After assuming that torsional buckling is ruled by warping and therefore neglecting the torsional stiffness, Kollár [1986] established the differential equations of such systems. The critical load is given as the smallest root of the cubic polynomial resulting from the determinant

$$\begin{vmatrix} (F_{cr}-\bar{F}_{cr,x}) & 0 & F_{cr}y_o \\ 0 & (F_{cr}-\bar{F}_{cr,y}) & -F_{cr}x_o \\ F_{cr}y_o & -F_{cr}x_o & i_p^2(F_{cr}-\bar{F}_{cr,\varphi}) \end{vmatrix} . \tag{7.84}$$

There is a formal resemblance between determinants (7.78) and (7.84). $F_{cr,x}$ and $F_{cr,y}$ in (7.78) refer to the Euler type critical load of the solid substitute cantilever where only the bending stiffness is taken into account. In a similar way, $F_{cr,\varphi}$ in (7.78) involves the bending stiffnesses of the elements - through their warping stiffness. The situation with $\bar{F}_{cr,x}$, $\bar{F}_{cr,y}$ and $\bar{F}_{cr,\varphi}$ in (7.84) is different. These critical loads refer to the critical loads of the coupled shear walls or, to be more precise to a substitute cantilever which is a set of coupled shear walls. With coupled shear walls, however, shear deformation plays a very important role. The calculation of $\bar{F}_{cr,x}$, $\bar{F}_{cr,y}$ and $\bar{F}_{cr,\varphi}$ therefore must be based on a method which also takes shear deformation into account. Such a method is presented in Chapter 5, where the in-plane stability analysis of frameworks and coupled shear walls is presented. The calculation of $\bar{F}_{cr,\varphi}$ is given in [Kollár, 1986].

If the system of coupled shear walls is subjected to uniformly distributed load, a similar procedure can be established. For solid support elements, Eqn (7.74) gives the critical load, making use of the part critical loads $N_{cr,x}$, $N_{cr,y}$ and $N_{cr,\varphi}$. If these part critical loads are again calculated by making use of shear deformation ($\bar{N}_{cr,x}$, $\bar{N}_{cr,y}$, $\bar{N}_{cr,\varphi}$), the resulting critical load is an approximation of the critical load of the system of coupled shear walls. This method, just like the one outlined above for the structures subjected to concentrated force, does not take into account the torsional stiffness of the elements.

7.7 CONCLUSIONS

1. Relying on the continuum method, the spatial stability analysis is easy and can be carried out in minutes for structures supported by walls and cores of open or closed cross-sections. To indicate how quickly the continuum method can be used and applied to practical problems, we only mention that the complete global stability analysis of the design example in Section 7.5.2.1, shown in Fig. 7-11, only took 13 minutes.

2. As a rule, a building approaches critical state in such a way that pure torsional buckling and flexural buckling in the two principal planes develop simultaneously. This interaction of the three modes results in an overall critical load which is smaller than either of the part critical loads. To avoid this unfavourable interaction, it is desirable to go for a doubly symmetric support system.

3. Pure torsional buckling plays a very important role. It is often smaller than the flexural critical loads. It follows that the overall critical load is often equal to the critical load of pure torsional buckling. This is the case with doubly symmetric support systems. With unsymmetric support systems, the overall critical load is even smaller than that of the pure torsional buckling. When the support system only has one axis of symmetry, the overall critical load is equal to or smaller than the critical load of pure torsional buckling, depending on which of the two flexural critical loads is smaller and which of them pure torsional buckling combines with.

4. Pure torsional buckling is characterized by two main factors: The torsional stiffness and the warping stiffness of the support system. The critical load of pure torsional buckling is a combination of the part critical load associated with the warping stiffness and the part critical load associated with the torsional stiffness. The value of both part critical loads depends on the radius of gyration of the ground plan of the building. This is a very important difference from the case of flexural buckling where the geometrical characteristics of the ground plan has no importance at all.

5. Of the two part critical loads for pure torsional buckling, the one associated with the warping stiffness is very important and often dominant. Apart from the geometrical characteristics of the ground plan,

its value also depends on the height of the building. The value of the overall critical load of a building can be increased most effectively by increasing the warping stiffness of the support system. The value of the warping stiffness can be increased in different ways. The best solution is achieved through geometrical considerations: The support elements should be positioned in such a way that they are as far from the shear centre as possible and that their bending stiffness is most effective against the rotation of the building around the shear centre.

6. Special attention should be paid to support systems with only torsional but no warping stiffness. This is often the case when an unstable column system is supported by a single core of closed - or partially closed - cross-section. The value of the critical load does not depend on either the height of the building or the nature of the load - two facts which should be kept in mind in structural design.

8

APPENDICES

8.1. ADDITIVE THEOREMS

Additive theorems are invaluable tools in dealing with complex problems in a simple way. As the procedures presented in this book often rely on additive theorems, a short summary is given below. A detailed analysis with the mathematical background and the limitation of application is given by Tarnai [1980, 1981] and is also available in Kollár's monograph [1991].

The principle of all additive theorems used in stability analysis is as follows. The complex problem is divided into simpler parts. These simpler problems are then solved and the critical load of each case is computed. The critical load of the original complex problem is finally obtained by producing the sum of the part problems. Summation can be direct or inverse addition. The great advantage of the additive theorems is that a simple solution to the part problems is usually available, whereas the solution of the original problem is often very difficult to produce.

8.1.1. Southwell's Theorem

Southwell [1922] first applied his theorem to vibration problems. Southwell's theorem applied to stability problems is defined as follows. If the stiffness of an elastic structure can be considered as the sum of the stiffnesses of several superimposed parts, then the sum of the critical loads of the component parts represents a lower bound to the actual critical load of the structure:

$$\lambda_o \geq \sum_{i=1}^{n} \lambda_i. \tag{8.1}$$

In inequality (8.1), λ_o is a lower bound to the smallest critical load of the structure and λ_i is the critical load of the structure when it only has its *i*-th part stiffness.

If the eigenvalue problem only has positive eigenvalues, Southwell's formula is conservative. The more similar the buckling shapes of the part problems are, the smaller error is made by the application of Southwell's formula. When the buckling shapes of the part problems which belong to the part stiffnesses are identical, equality in formula (8.1) holds, i.e., λ_o is the exact critical load of the problem.

8.1.2. Dunkerley's Theorem

Dunkerley [1894] first used his formula for the vibration problem of shafts. The Dunkerley theorem applied to stability problems is defined as follows. If the load system of an elastic structure can be considered as the sum of individual loads, then by the inverse addition of the critical loads which belong to the individual loads, a lower bound of the critical load of the structure is obtained:

$$\frac{1}{\lambda_o} \leq \sum_{i=1}^{n} \frac{1}{\lambda_i} . \tag{8.2}$$

In inequality (8.2), λ_o is a lower bound to the smallest critical load of the structure and λ_i is the critical load of the structure when it is only subjected to the *i*-th load of the load system.

The critical load obtained by the application of Dunkerley's formula is always conservative. What was said about the accuracy of the Southwell formula above, holds for Dunkerley's theorem as well.

8.1.3. The Föppl-Papkovics Theorem

Föppl [1933] developed a theory for the stability analysis of elastic structures, based on the partial stiffening of its elements. The principle of his theorem is as follows. The elastic system is characterized by *n* stiffness characteristics. The system is first considered with the value of all the stiffness parameters increased to infinity, except the *i*-th one, which is left untouched. Critical load parameter λ_i of this system is computed. The procedure is then repeated and always another stiffness parameter is untouched while the value of all the others is assumed to be infinitely great. The critical load of the original system λ_o is obtained from the reciprocal formula

$$\frac{1}{\lambda_o} \cong \sum_{i=1}^{n} \frac{1}{\lambda_i} . \tag{8.3}$$

If all the deformations are independent of each other and the load is of first type (which does work on all kinematically possible deformations), then the approximate formula is conservative [Papkovics,

1963], i.e.,

$$\frac{1}{\lambda_o} \leq \sum_{i=1}^{n} \frac{1}{\lambda_i} . \tag{8.4}$$

Equality in formula (8.4) holds, if the buckling shapes are identical.

8.1.4. The Rankine Formula

Analysing axially compressed columns, Gordon developed an approximate formula for the elasto-plastic critical load, based on Hodgkinson's [1840] experimental results. Rankine [1863] applied Gordon's formula to columns with pinned supports and arrived at

$$\frac{1}{F^{ep}_{cr}} \cong \frac{1}{F^{p}_{cr}} + \frac{1}{F^{e}_{cr}} . \tag{8.5}$$

In formula (8.5),

F^{ep}_{cr} is the elasto-plastic critical load,

F^{e}_{cr} is the elastic critical load,

F^{p}_{cr} is the rigid-plastic collapse load.

Merchant [1954] extended Rankine's formula to the elasto-plastic analysis of frameworks.

The Rankine formula has proved to offer good approximate solution for individual columns and for frameworks. However, no mathematical proof is yet available whether or not the formula is conservative.

8.2 DERIVATION OF THE GOVERNING EQUATION FOR TORSIONAL BUCKLING

Consider a column with a fixed lower end and a free upper end, having an unsymmetrical, thin-walled cross-section. The column and its cross-section is shown in Fig. 7-6 in Chapter 7. The column is subjected to a uniformly distributed normal load of intensity q. The coordinate system is set in such a way that axis z coincides with the centroidal axis of the column and axes x and y are the principal axes of inertia of the free upper cross-section. Axes x and y are fixed to the upper cross-section. They move together with the cross-section and remain principal axes of inertia during buckling. Point C denotes the centroid of the cross-section and x_o and y_o are the coordinates of shear centre O.

The cross-sections undergo translation and rotation during buckling. The translation is defined by displacements u and v of shear centre O in directions x and y, respectively. During translation, shear centre O moves to O' and centroid C to C'. The rotation of the cross-section about the shear centre is denoted by angle φ and the final position of the centroid is C'' (Fig. 7-6/b). The final translations of the centroid are

$$u + y_o\varphi \tag{8.6}$$

and

$$v - x_o\varphi \; . \tag{8.7}$$

The differential equation of the buckling of the shear centre axis in plane *xz* is established first and then the differential equation of buckling in plane *yz* is given disregarding the details.

Consider an elementary section of the shear centre axis (Fig. 8-1). The following equilibrium equations can be established. Equilibrium of moments:

$$N(\mathrm{d}u + y_o\mathrm{d}\varphi) - \mathrm{d}M + T\mathrm{d}z = 0 \; . \tag{8.8}$$

Hence:

$$T = - N(u' + y_o\varphi') + M' \; . \tag{8.9}$$

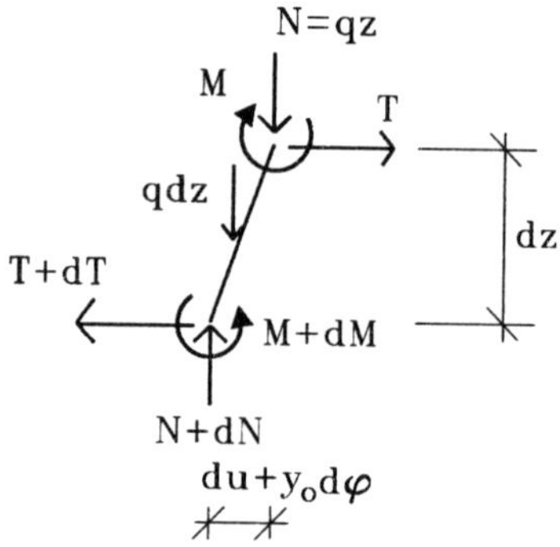

Fig. 8-1. Elementary section of the column

Equilibrium of horizontal forces:

$$\mathrm{d}T = 0 \; ,$$

or

$$T' = 0 \; . \tag{8.10}$$

Equilibrium of vertical forces:

$$q\mathrm{d}z - \mathrm{d}N = 0 \; . \tag{8.11}$$

Hence:

$$N' = q \ . \tag{8.12}$$

After differentiating Eqn (8.9) once with respect to z and making use of Eqs (8.10) and (8.12), the differential equation

$$-q(u' + y_o\varphi') - qz(u'' + y_o\varphi'') + M'' = 0 \tag{8.13}$$

is obtained. According to elementary stress analysis, the equation

$$M = -EIu''$$

holds, so that the differential equation for the buckling of the shear centre axis can be rewritten as

$$u'''' + \frac{N}{EI_y}(u'' + y_o\varphi'') + \frac{q}{EI_y}(u' + y_o\varphi') = 0 \ . \tag{8.14}$$

A similar derivation results in the differential equation of buckling in plane yz as

$$v'''' + \frac{N}{EI_x}(v'' - x_o\varphi'') + \frac{q}{EI_x}(v' - x_o\varphi') = 0 \ . \tag{8.15}$$

When the origin of the coordinate system is fixed on top of the column, the boundary conditions are obtained as follows.

a) The displacement in directions x and y is zero at the top of the column:

$$u(0) = v(0) = 0 \ . \tag{8.16}$$

b) The tangent to the column is parallel to axis z at the bottom of the column:

$$u'(H) = v'(H) = 0 \ . \tag{8.17}$$

c) No bending moment develops at the top of the column:

$$u''(0) = v''(0) = 0 \ . \tag{8.18}$$

d) The shear force vanishes at the top of the column:

$$u'''(0) = v'''(0) = 0 \ . \tag{8.19}$$

The two differential equations responsible for the flexural buckling of the column [Eqs (8.14) and (8.15)] contain the three unknown quantities u, v and φ. The third equation needed to control torsional-flexural-buckling is obtained by considering the torsion of the column. This procedure follows now.

Consider a longitudinal strip of cross-section tds defined by coordinates x and y in the plane of the cross-section of the column (Fig. 7-6/b in Chapter 7 and Fig. 8-2/a).

The components of the translation of cross-section tds at z assume the value

$$u + (y_o - y)\varphi \tag{8.20}$$

and

$$v - (x_o - x)\varphi \ . \tag{8.21}$$

The elementary section of length dz of the longitudinal strip is subjected to a compressive unit stress

$$\sigma = \frac{zq}{A} \ , \tag{8.22}$$

where A is the cross-sectional area of the longitudinal strip considered to be an individual bar. The compressive force acting on the slightly rotated end of the element defined by coordinate z (Fig. 8-2/b) is

$$N = \sigma t ds = zq \frac{tds}{A} \ . \tag{8.23}$$

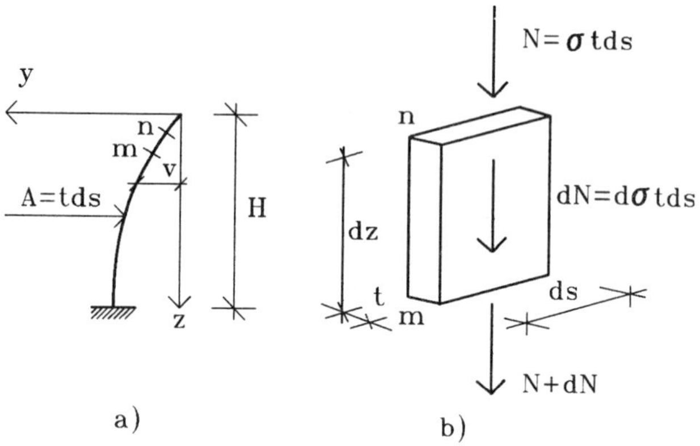

Fig. 8-2. Cantilever of cross-section *tds*: a) longitudinal strip and b) elementary section

The equilibrium of the element of length dz in plane yz is expressed by the following equations.

Equilibrium of moments:

$$N\left(dv - (x_o - x)d\varphi\right) - dM + Tdz = 0 \ .$$

Hence:

$$T = M' - N\left(v' - (x - x_o)\varphi'\right) . \tag{8.24}$$

Equilibrium of horizontal forces:

$$\mathrm{d}T = 0$$

or

$$T' = 0 . \tag{8.25}$$

Equilibrium of vertical forces:

$$q\mathrm{d}z\frac{t\mathrm{d}s}{A} - \mathrm{d}N = 0 .$$

Hence:

$$N' = q\frac{t\mathrm{d}s}{A} . \tag{8.26}$$

After differentiating Eqn (8.24) once, substituting it into Eqn (8.25) and making use of Eqn (8.26), we obtain

$$M'' - qz\frac{t\mathrm{d}s}{A}\left(v'' - (x_o - x)\varphi''\right) - q\frac{t\mathrm{d}s}{A}\left(v' - (x_o - x)\varphi'\right) = 0 . \tag{8.27}$$

Introducing

$$M = -EIv''$$

from elementary stress analysis into Eqn (8.27), we arrive at the differential equation of buckling in plane yz as

$$EI_x v'''' = -qz\frac{t\mathrm{d}s}{A}\left(v'' - (x_o - x)\varphi''\right) - q\frac{t\mathrm{d}s}{A}\left(v' - (x_o - x)\varphi'\right) . \tag{8.28}$$

Equation (8.28) assumes the same form as that of the governing differential equation of an ordinary beam subjected to bending. With that beam, the right-hand side of the equation represents the intensity of the load, perpendicular to the axis of the beam. In our case, the right-hand side of the equation can be considered as a fictitious load intensity in plane yz:

$$-qz\frac{t\mathrm{d}s}{A}\left(v'' - (x_o - x)\varphi''\right) - q\frac{t\mathrm{d}s}{A}\left(v' - (x_o - x)\varphi'\right) . \tag{8.29}$$

Buckling also develops in plane xz. The intensity of the fictitious load in that plane assumes the form

$$-qz\frac{t\mathrm{d}s}{A}\left(u'' + (y_o - y)\varphi''\right) - q\frac{t\mathrm{d}s}{A}\left(u' + (y_o - y)\varphi'\right) . \tag{8.30}$$

These fictitious loads give rise to twisting. Taking the moment of the above forces about the shear centre axis for one longitudinal strip, the following torque per unit length is obtained:

$$\mathrm{d}m_z = \frac{qz\,t\,\mathrm{d}s}{A}(x_o - x)\left(v'' - (x_o - x)\varphi''\right) + \frac{qt\,\mathrm{d}s}{A}(x_o - x)\left(v' - (x_o - x)\varphi'\right) -$$

$$- \frac{qz\,t\,\mathrm{d}s}{A}(y_o - y)\left(u'' + (y_o - y)\varphi''\right) - \frac{qt\,\mathrm{d}s}{A}(y_o - y)\left(u' + (y_o - y)\varphi'\right). \qquad (8.31)$$

After integrating over the entire cross-section and taking into account that the expressions

$$\int_A t\mathrm{d}s = A,$$

$$\int_A x^2 t\mathrm{d}s = I_y, \qquad \int_A y^2 t\mathrm{d}s = I_x,$$

$$I_o = I_x + I_y + A(x_o^2 + y_o^2) \qquad (8.32)$$

and

$$\int_A xt\mathrm{d}s = \int_A yt\mathrm{d}s = 0$$

hold, the equation

$$m_z = \int \mathrm{d}m_z = qz(x_o v'' - y_o u'') + q(x_o v' - y_o u') - qz\frac{I_o}{A}\varphi'' - q\frac{I_o}{A}\varphi' \qquad (8.33)$$

is obtained. In the above expressions

I_x and I_y are the second moment of area of the cross-section,

I_o is the polar moment of inertia about the shear centre.

The torque per unit length of the internal forces for non-uniform torsion is given by the following expression

$$\frac{\mathrm{d}M_t}{\mathrm{d}z} = GJ\varphi'' - EI_\omega\varphi'''' . \qquad (8.34)$$

In equation (8.34)

G is the modulus of elasticity in shear,

J is the torsional constant,

GJ is the torsional stiffness,
I_ω is the warping constant,
EI_ω is the warping stiffness.

The positive direction of M_t and m_z is given by the right-hand rule (Fig. 8-3):

$$m_z = - \frac{dM_t}{dz} . \tag{8.35}$$

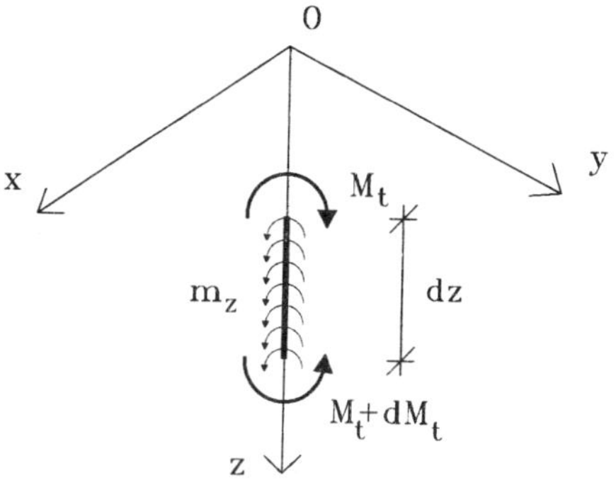

Fig. 8-3. Sign convention for torques

After substituting formulae (8.34) and (8.35) into Eqn (8.33), we finally arrive at

$$EI_\omega \varphi'''' - \left(GJ - qz\frac{I_o}{A}\right)\varphi'' + q\frac{I_o}{A}\varphi' - qz(x_o v'' - y_o u'') -$$

$$- q(x_o v' - y_o u') = 0, \tag{8.36}$$

or in a shorter form

$$EI_\omega \varphi'''' - \left(GJ\varphi' - qzi_p^2\varphi'\right)' - \left(qz(x_o v' - y_o u')\right)' = 0. \tag{8.37}$$

In Eqn (8.37)

$$i_p = \sqrt{\frac{I_o}{A}} \tag{8.38}$$

is the radius of gyration.

Equation (8.37) is the fourth-order differential equation of variable coefficients of pure torsional buckling, which was first derived by

Vlasov [1961] when he investigated the spatial stability of a beam with thin-walled cross-section. The equation can be used for the stability analysis of buildings supported by cores of thin-walled cross-section. For such analysis, EI_ω and GJ are the warping and torsional stiffnesses of the support system, respectively. The calculation of the radius of gyration, however, must be based on the geometrical characteristics of the ground plan of the building. This follows from the derivation of the fictitious load, which causes the twist of the system. The fictitious load originates from the vertical load of the system, which is transmitted by the longitudinal fibres of the cross-section. With buildings, the columns and the cores together are considered to be the longitudinal fibres transmitting the vertical load. Assuming that they are uniformly distributed over the ground plan, we arrive at the radius of gyration based on the ground plan of the building. With buildings of rectangular ground plan the radius of gyration assumes the form

$$i_p = \sqrt{\frac{(L^2 + B^2)}{12} + t^2} \ . \tag{8.39}$$

In formula (8.39), L and B are size of the ground plan and t is the distance between the centroid of the ground plan and the shear centre of the support system.

Bearing in mind that axes x and y of the coordinate system are fixed at the top of the column, the boundary conditions are as follows:

a) The coordinate system at the top of the column rotates together with the cross-section so that no torsion can be measured there:

$$\varphi(0) = 0 \ . \tag{8.40}$$

b) Because of the clamped lower end of the column, there is no warping at the bottom of the column:

$$\varphi'(H) = 0 \ . \tag{8.41}$$

c) No compressive stresses act at the top of the column so warping stresses cannot develop either:

$$\varphi''(0) = 0 \ . \tag{8.42}$$

d) The torsional moment equals zero at the top of the column:

$$GJ\varphi'(0) - EI_\omega\varphi'''(0) = 0 \ . \tag{8.43}$$

Eqs (8.14), (8.15) and (8.37) are the three simultaneous differential equations for buckling by bending and torsion. The smallest eigenvalue of the problem is the critical load parameter q_{cr}.

The order of magnitude of differential equation (8.37) can be reduced. After carrying out the integration of Eqn (8.37), the differential equation

$$\varphi''' - \frac{GJ}{EI_\omega}\varphi' + \frac{qI_o}{EI_\omega A}\int_A z\varphi''\mathrm{d}z + \frac{qI_o}{A}\varphi - \frac{qx_o}{EI_\omega}\int zv''\mathrm{d}z +$$

$$+ \frac{qy_o}{EI_\omega}\int zu''\mathrm{d}z - \frac{qx_o}{EI_\omega}v + \frac{qy_o}{EI_\omega}u + c_1 = 0\ , \qquad (8.44)$$

is obtained. Substituting

$$\int z\varphi''\mathrm{d}z = z\varphi' - \varphi + c_2\ ,$$

$$\int zv''\mathrm{d}z = zv' - v + c_3\ ,$$

and

$$\int zu''\mathrm{d}z = zu' - u + c_4$$

into Eqn (8.43), we obtain

$$\varphi''' - \left(\frac{GJ}{EI_\omega} - \frac{qI_o}{EI_\omega A}z\right)\varphi' - \frac{q}{EI_\omega}z\left(x_o v' - y_o u'\right) + c_5 = 0 \qquad (8.45)$$

Boundary condition (8.43) yields zero for the value of the constant of integration c_5. Thus, since the order of Eqs (8.14) and (8.15) can be reduced in a similar way, the simultaneous differential equation for buckling by bending and torsion can be rewritten as

$$u''' + \frac{qz}{EI_y}\left(u' + y_o\varphi'\right) = 0\ , \qquad (8.46)$$

$$v''' + \frac{qz}{EI_x}\left(v' - x_o\varphi'\right) = 0\ , \qquad (8.47)$$

$$\varphi''' - \left(\frac{GJ}{EI_\omega} - \frac{qi_p^2}{EI_\omega}z\right)\varphi' - \frac{q}{EI_\omega}z\left(x_o v' - y_o u'\right) = 0\ . \qquad (8.48)$$

The boundary conditions for the differential equations are

$$u(0) = v(0) = \varphi(0) = 0\ , \qquad (8.49)$$

$$u'(H) = v'(H) = \varphi'(H) = 0\ , \qquad (8.50)$$

$$u''(0) = v''(0) = \varphi''(0) = 0\ . \qquad (8.51)$$

8.3 TABLES FOR TORSIONAL BUCKLING

Table 8-1. Load parameter λ for $\tau_x = 0.1$

$\frac{N_{cr,\varphi}}{N_{cr,y}}$ \ $\frac{N_{cr,\varphi}}{N_{cr,x}}$	0.2	0.4	0.6	0.8	1.0	2.0	3.0	4.0	5.0	τ_y
0.0	1.000	1.000	1.000	1.000	1.000	0.500	0.333	0.250	0.200	-
0.2	1.000	1.000	1.000	1.000	1.000	0.500	0.333	0.250	0.200	0.0
	0.991	0.976	0.950	0.905	0.833	0.481	0.326	0.247	0.198	0.2
	0.964	0.916	0.857	0.787	0.715	0.443	0.311	0.238	0.192	0.4
	0.926	0.846	0.767	0.692	0.626	0.402	0.290	0.226	0.184	0.6
	0.881	0.777	0.690	0.617	0.556	0.365	0.269	0.212	0.176	0.8
	0.835	0.716	0.626	0.557	0.501	0.333	0.250	0.201	0.167	1.0
0.4	0.996	0.996	0.996	0.996	0.996	0.500	0.333	0.250	0.200	0.0
	0.987	0.972	0.947	0.904	0.833	0.481	0.326	0.247	0.198	0.2
	0.960	0.914	0.854	0.786	0.714	0.443	0.310	0.238	0.192	0.4
	0.922	0.844	0.765	0.691	0.625	0.402	0.290	0.226	0.184	0.6
	0.878	0.776	0.689	0.616	0.556	0.365	0.269	0.212	0.176	0.8
	0.833	0.715	0.626	0.557	0.501	0.333	0.250	0.201	0.167	1.0
0.6	0.988	0.988	0.988	0.988	0.988	0.500	0.333	0.250	0.200	0.0
	0.980	0.965	0.942	0.900	0.831	0.481	0.326	0.247	0.198	0.2
	0.954	0.909	0.852	0.785	0.713	0.443	0.310	0.238	0.192	0.4
	0.918	0.841	0.764	0.691	0.625	0.402	0.290	0.226	0.184	0.6
	0.874	0.774	0.688	0.616	0.555	0.365	0.269	0.212	0.176	0.8
	0.830	0.714	0.626	0.556	0.501	0.333	0.250	0.201	0.167	1.0
0.8	0.970	0.970	0.970	0.970	0.970	0.499	0.333	0.250	0.200	0.0
	0.962	0.950	0.929	0.891	0.827	0.481	0.326	0.247	0.198	0.2
	0.939	0.899	0.846	0.781	0.712	0.443	0.311	0.238	0.192	0.4
	0.907	0.835	0.760	0.689	0.624	0.402	0.290	0.226	0.184	0.6
	0.867	0.770	0.687	0.615	0.555	0.365	0.269	0.212	0.176	0.8
	0.825	0.712	0.624	0.556	0.501	0.334	0.250	0.201	0.167	1.0
1.0	0.910	0.910	0.910	0.910	0.910	0.499	0.333	0.250	0.200	0.0
	0.906	0.901	0.890	0.867	0.818	0.481	0.326	0.247	0.198	0.2
	0.895	0.870	0.830	0.774	0.708	0.443	0.311	0.238	0.192	0.4
	0.874	0.821	0.755	0.687	0.622	0.402	0.290	0.226	0.184	0.6
	0.846	0.764	0.684	0.614	0.554	0.365	0.269	0.212	0.175	0.8
	0.812	0.708	0.622	0.555	0.500	0.334	0.250	0.201	0.167	1.0
2.0	0.495	0.495	0.495	0.495	0.495	0.495	0.333	0.250	0.200	0.0
	0.495	0.495	0.495	0.495	0.495	0.478	0.326	0.247	0.198	0.2
	0.495	0.495	0.495	0.494	0.494	0.440	0.310	0.238	0.192	0.4
	0.495	0.495	0.494	0.493	0.492	0.401	0.290	0.226	0.184	0.6
	0.495	0.494	0.493	0.491	0.488	0.364	0.269	0.212	0.175	0.8
	0.494	0.493	0.491	0.488	0.477	0.333	0.250	0.201	0.167	1.0

to be continued

Table 8-1. Continued. Load parameter λ for $\tau_x = 0.1$.

$\frac{N_{cr,\varphi}}{N_{cr,y}}$ \ $\frac{N_{cr,\varphi}}{N_{cr,x}}$	0.2	0.4	0.6	0.8	1.0	2.0	3.0	4.0	5.0	τ_y
	0.331	0.331	0.331	0.331	0.331	0.331	0.331	0.250	0.200	0.0
	0.331	0.331	0.331	0.331	0.331	0.331	0.325	0.247	0.198	0.2
3.0	0.331	0.331	0.331	0.331	0.331	0.330	0.309	0.238	0.192	0.4
	0.331	0.331	0.331	0.331	0.331	0.329	0.289	0.226	0.184	0.6
	0.331	0.331	0.331	0.331	0.331	0.325	0.269	0.212	0.175	0.8
	0.331	0.331	0.331	0.331	0.331	0.316	0.250	0.200	0.167	1.0
	0.249	0.249	0.249	0.249	0.249	0.249	0.249	0.249	0.200	0.0
	0.249	0.249	0.249	0.249	0.249	0.249	0.249	0.246	0.198	0.2
4.0	0.249	0.249	0.249	0.249	0.249	0.249	0.249	0.237	0.192	0.4
	0.249	0.249	0.249	0.249	0.249	0.249	0.248	0.226	0.184	0.6
	0.249	0.249	0.249	0.249	0.249	0.249	0.247	0.212	0.175	0.8
	0.249	0.249	0.249	0.249	0.249	0.248	0.244	0.200	0.167	1.0
	0.199	0.199	0.199	0.199	0.199	0.199	0.199	0.199	0.199	0.0
	0.199	0.199	0.199	0.199	0.199	0.199	0.199	0.199	0.197	0.2
5.0	0.199	0.199	0.199	0.199	0.199	0.199	0.199	0.199	0.191	0.4
	0.199	0.199	0.199	0.199	0.199	0.199	0.199	0.198	0.184	0.6
	0.199	0.199	0.199	0.199	0.199	0.199	0.199	0.198	0.175	0.8
	0.199	0.199	0.199	0.199	0.199	0.199	0.198	0.196	0.167	1.0

Table 8-2. Load parameter λ for $\tau_x = 0.2$.

$\frac{N_{cr,\varphi}}{N_{cr,y}}$ \ $\frac{N_{cr,\varphi}}{N_{cr,x}}$	0.2	0.4	0.6	0.8	1.0	2.0	3.0	4.0	5.0	τ_y
0.0	1.000	1.000	1.000	1.000	1.000	0.500	0.333	0.250	0.200	-
	0.993	0.993	0.993	0.993	0.993	0.500	0.333	0.250	0.200	0.0
	0.984	0.969	0.945	0.902	0.832	0.481	0.326	0.247	0.198	0.2
0.2	0.957	0.912	0.853	0.785	0.713	0.443	0.311	0.238	0.192	0.4
	0.920	0.842	0.764	0.691	0.625	0.402	0.290	0.226	0.184	0.6
	0.876	0.774	0.688	0.616	0.555	0.365	0.269	0.212	0.176	0.8
	0.832	0.714	0.626	0.556	0.501	0.333	0.250	0.201	0.167	1.0
	0.978	0.978	0.978	0.978	0.978	0.500	0.333	0.250	0.200	0.0
	0.969	0.956	0.933	0.894	0.827	0.481	0.326	0.247	0.198	0.2
0.4	0.945	0.902	0.846	0.781	0.711	0.443	0.310	0.238	0.192	0.4
	0.909	0.836	0.760	0.688	0.623	0.402	0.290	0.226	0.184	0.6
	0.868	0.770	0.685	0.614	0.554	0.365	0.269	0.212	0.176	0.8
	0.825	0.711	0.623	0.555	0.500	0.333	0.250	0.201	0.167	1.0

to be continued

Table 8-2. Continued. Load parameter λ for $\tau_x = 0.2$.

$\frac{N_{cr,\varphi}}{N_{cr,y}}$ \ $\frac{N_{cr,\varphi}}{N_{cr,x}}$	0.2	0.4	0.6	0.8	1.0	2.0	3.0	4.0	5.0	τ_y
	0.952	0.952	0.952	0.952	0.952	0.500	0.333	0.250	0.200	0.0
	0.945	0.933	0.914	0.880	0.820	0.481	0.326	0.247	0.198	0.2
0.6	0.923	0.885	0.835	0.774	0.708	0.443	0.310	0.238	0.192	0.4
	0.892	0.825	0.753	0.685	0.621	0.402	0.290	0.226	0.184	0.6
	0.854	0.763	0.681	0.612	0.553	0.365	0.269	0.212	0.175	0.8
	0.815	0.706	0.621	0.553	0.499	0.333	0.250	0.201	0.167	1.0
	0.907	0.907	0.907	0.907	0.907	0.500	0.333	0.250	0.200	0.0
	0.902	0.894	0.890	0.855	0.808	0.481	0.326	0.247	0.198	0.2
0.8	0.886	0.856	0.816	0.763	0.702	0.443	0.310	0.238	0.192	0.4
	0.862	0.806	0.743	0.679	0.618	0.402	0.290	0.226	0.184	0.6
	0.831	0.751	0.676	0.609	0.551	0.364	0.269	0.212	0.175	0.8
	0.797	0.698	0.617	0.552	0.498	0.333	0.250	0.201	0.167	1.0
	0.834	0.834	0.834	0.834	0.834	0.500	0.333	0.250	0.200	0.0
	0.832	0.827	0.820	0.808	0.781	0.481	0.326	0.247	0.198	0.2
1.0	0.823	0.806	0.781	0.743	0.691	0.442	0.310	0.238	0.192	0.4
	0.809	0.773	0.725	0.670	0.613	0.402	0.290	0.226	0.184	0.6
	0.789	0.730	0.666	0.604	0.549	0.364	0.269	0.212	0.175	0.8
	0.766	0.685	0.611	0.549	0.496	0.333	0.250	0.201	0.167	1.0
	0.481	0.481	0.481	0.481	0.481	0.481	0.333	0.250	0.200	0.0
	0.481	0.481	0.481	0.481	0.481	0.467	0.326	0.247	0.198	0.2
2.0	0.481	0.481	0.481	0.480	0.479	0.433	0.310	0.237	0.192	0.4
	0.481	0.480	0.479	0.478	0.475	0.396	0.289	0.226	0.184	0.6
	0.481	0.479	0.477	0.474	0.467	0.361	0.269	0.212	0.175	0.8
	0.480	0.478	0.473	0.466	0.453	0.331	0.250	0.200	0.167	1.0
	0.326	0.326	0.326	0.326	0.326	0.326	0.326	0.250	0.200	0.0
	0.326	0.326	0.326	0.326	0.326	0.326	0.321	0.247	0.198	0.2
3.0	0.326	0.326	0.326	0.326	0.326	0.326	0.306	0.237	0.192	0.4
	0.326	0.326	0.326	0.326	0.326	0.324	0.287	0.226	0.184	0.6
	0.326	0.326	0.326	0.326	0.326	0.320	0.267	0.212	0.175	0.8
	0.326	0.326	0.326	0.326	0.325	0.311	0.249	0.200	0.167	1.0
	0.247	0.247	0.247	0.247	0.247	0.247	0.247	0.247	0.200	0.0
	0.247	0.247	0.247	0.247	0.247	0.247	0.247	0.243	0.198	0.2
4.0	0.247	0.247	0.247	0.247	0.247	0.247	0.246	0.235	0.192	0.4
	0.247	0.247	0.247	0.247	0.247	0.246	0.245	0.224	0.184	0.6
	0.247	0.247	0.247	0.247	0.246	0.246	0.243	0.211	0.175	0.8
	0.247	0.247	0.247	0.246	0.246	0.245	0.238	0.199	0.167	1.0
	0.198	0.198	0.198	0.198	0.198	0.198	0.198	0.198	0.198	0.0
	0.198	0.198	0.198	0.198	0.198	0.198	0.198	0.198	0.195	0.2
5.0	0.198	0.198	0.198	0.198	0.198	0.198	0.198	0.197	0.191	0.4
	0.198	0.198	0.198	0.198	0.198	0.198	0.198	0.197	0.183	0.6
	0.198	0.198	0.198	0.198	0.198	0.198	0.197	0.195	0.174	0.8
	0.198	0.198	0.198	0.198	0.198	0.197	0.197	0.192	0.166	1.0

Table 8-3. Load parameter λ for $\tau_x = 0.3$

$\frac{N_{cr,\varphi}}{N_{cr,y}}$ \ $\frac{N_{cr,\varphi}}{N_{cr,x}}$	0.2	0.4	0.6	0.8	1.0	2.0	3.0	4.0	5.0	τ_y
0.0	1.000	1.000	1.000	1.000	1.000	0.500	0.333	0.250	0.200	-
0.2	0.981	0.981	0.981	0.981	0.981	0.500	0.333	0.250	0.200	0.0
	0.973	0.959	0.936	0.895	0.828	0.481	0.326	0.247	0.198	0.2
	0.947	0.903	0.847	0.781	0.712	0.443	0.311	0.238	0.192	0.4
	0.911	0.836	0.760	0.688	0.623	0.402	0.290	0.226	0.184	0.6
	0.870	0.771	0.686	0.615	0.554	0.365	0.269	0.212	0.176	0.8
	0.826	0.711	0.623	0.555	0.500	0.333	0.250	0.201	0.167	1.0
0.4	0.950	0.950	0.950	0.950	0.950	0.500	0.333	0.250	0.200	0.0
	0.943	0.931	0.912	0.878	0.819	0.481	0.326	0.247	0.198	0.2
	0.922	0.883	0.833	0.772	0.706	0.443	0.310	0.238	0.192	0.4
	0.889	0.822	0.751	0.683	0.620	0.402	0.290	0.226	0.184	0.6
	0.851	0.760	0.680	0.611	0.552	0.365	0.269	0.212	0.176	0.8
	0.812	0.704	0.619	0.553	0.498	0.333	0.250	0.201	0.167	1.0
0.6	0.906	0.906	0.906	0.906	0.906	0.500	0.333	0.250	0.200	0.0
	0.900	0.891	0.877	0.850	0.804	0.481	0.326	0.247	0.198	0.2
	0.883	0.852	0.811	0.758	0.698	0.442	0.310	0.238	0.192	0.4
	0.857	0.801	0.739	0.675	0.616	0.401	0.290	0.226	0.184	0.6
	0.826	0.746	0.671	0.606	0.550	0.365	0.269	0.212	0.175	0.8
	0.791	0.694	0.614	0.550	0.496	0.333	0.250	0.200	0.167	1.0
0.8	0.844	0.844	0.844	0.844	0.844	0.499	0.333	0.250	0.200	0.0
	0.840	0.835	0.826	0.809	0.778	0.481	0.326	0.247	0.198	0.2
	0.829	0.808	0.778	0.737	0.687	0.442	0.310	0.238	0.192	0.4
	0.811	0.769	0.719	0.663	0.609	0.401	0.290	0.226	0.184	0.6
	0.788	0.724	0.660	0.599	0.546	0.363	0.269	0.212	0.175	0.8
	0.761	0.679	0.606	0.545	0.494	0.333	0.250	0.200	0.167	1.0
1.0	0.770	0.770	0.770	0.770	0.770	0.499	0.333	0.250	0.200	0.0
	0.768	0.765	0.760	0.752	0.736	0.481	0.326	0.247	0.198	0.2
	0.761	0.750	0.732	0.705	0.667	0.441	0.310	0.237	0.192	0.4
	0.751	0.725	0.689	0.646	0.599	0.400	0.290	0.226	0.184	0.6
	0.736	0.692	0.642	0.589	0.540	0.363	0.269	0.212	0.175	0.8
	0.718	0.657	0.595	0.539	0.491	0.332	0.250	0.200	0.167	1.0
2.0	0.464	0.464	0.464	0.464	0.464	0.464	0.333	0.250	0.200	0.0
	0.464	0.464	0.463	0.463	0.463	0.451	0.326	0.247	0.198	0.2
	0.463	0.463	0.462	0.461	0.460	0.422	0.309	0.237	0.192	0.4
	0.463	0.462	0.460	0.457	0.454	0.388	0.288	0.226	0.184	0.6
	0.462	0.460	0.457	0.452	0.445	0.356	0.267	0.212	0.175	0.8
	0.460	0.457	0.451	0.443	0.432	0.328	0.249	0.200	0.167	1.0

to be continued

Table 8-3. Continued. Load parameter λ for $\tau_x = 0.3$.

$\frac{N_{cr,\varphi}}{N_{cr,y}}$ \ $\frac{N_{cr,\varphi}}{N_{cr,x}}$	0.2	0.4	0.6	0.8	1.0	2.0	3.0	4.0	5.0	τ_y
3.0	0.319	0.319	0.319	0.319	0.319	0.319	0.319	0.250	0.200	0.0
	0.319	0.319	0.319	0.319	0.319	0.319	0.314	0.246	0.198	0.2
	0.319	0.319	0.319	0.319	0.319	0.318	0.301	0.237	0.192	0.4
	0.319	0.319	0.319	0.319	0.319	0.316	0.283	0.225	0.184	0.6
	0.319	0.319	0.319	0.318	0.318	0.309	0.264	0.212	0.175	0.8
	0.319	0.319	0.318	0.318	0.316	0.300	0.247	0.199	0.167	1.0
4.0	0.243	0.243	0.243	0.243	0.243	0.243	0.243	0.243	0.200	0.0
	0.243	0.243	0.243	0.243	0.243	0.243	0.243	0.240	0.198	0.2
	0.243	0.243	0.243	0.243	0.243	0.242	0.242	0.232	0.191	0.4
	0.243	0.243	0.243	0.243	0.242	0.242	0.240	0.221	0.184	0.6
	0.243	0.243	0.243	0.242	0.242	0.241	0.236	0.209	0.174	0.8
	0.243	0.243	0.242	0.242	0.242	0.240	0.231	0.198	0.167	1.0
5.0	0.195	0.195	0.195	0.195	0.195	0.195	0.195	0.195	0.195	0.0
	0.195	0.195	0.195	0.195	0.195	0.195	0.195	0.195	0.194	0.2
	0.195	0.195	0.195	0.195	0.195	0.195	0.195	0.195	0.188	0.4
	0.195	0.195	0.195	0.195	0.195	0.195	0.195	0.194	0.181	0.6
	0.195	0.195	0.195	0.195	0.195	0.194	0.194	0.191	0.173	0.8
	0.195	0.195	0.195	0.195	0.195	0.194	0.193	0.188	0.165	1.0

Table 8-4. Load parameter λ for $\tau_x = 0.4$

$\frac{N_{cr,\varphi}}{N_{cr,y}}$ \ $\frac{N_{cr,\varphi}}{N_{cr,x}}$	0.2	0.4	0.6	0.8	1.0	2.0	3.0	4.0	5.0	τ_y
0.0	1.000	1.000	1.000	1.000	1.000	0.500	0.333	0.250	0.200	-
0.2	0.966	0.966	0.966	0.966	0.966	0.500	0.333	0.250	0.200	0.0
	0.957	0.945	0.923	0.886	0.823	0.481	0.326	0.247	0.198	0.2
	0.934	0.893	0.840	0.776	0.708	0.443	0.310	0.238	0.192	0.4
	0.900	0.829	0.756	0.685	0.621	0.402	0.290	0.226	0.184	0.6
	0.860	0.765	0.682	0.612	0.553	0.365	0.269	0.212	0.175	0.8
	0.819	0.707	0.621	0.553	0.499	0.333	0.250	0.201	0.167	1.0
0.4	0.918	0.918	0.918	0.918	0.918	0.500	0.333	0.250	0.200	0.0
	0.912	0.902	0.885	0.857	0.806	0.481	0.326	0.247	0.198	0.2
	0.893	0.859	0.815	0.760	0.699	0.442	0.310	0.238	0.192	0.4
	0.864	0.805	0.740	0.676	0.616	0.401	0.290	0.226	0.184	0.6
	0.831	0.747	0.672	0.606	0.549	0.364	0.269	0.212	0.175	0.8
	0.795	0.695	0.614	0.550	0.496	0.333	0.250	0.201	0.167	1.0

to be continued

Table 8-4. Continued. Load parameter λ for $\tau_x = 0.4$.

$\frac{N_{cr,\varphi}}{N_{cr,y}}$ \ $\frac{N_{cr,\varphi}}{N_{cr,x}}$	0.2	0.4	0.6	0.8	1.0	2.0	3.0	4.0	5.0	τ_y
0.6	0.857	0.857	0.857	0.857	0.857	0.500	0.333	0.250	0.200	0.0
	0.853	0.846	0.835	0.816	0.781	0.481	0.326	0.247	0.198	0.2
	0.840	0.815	0.781	0.738	0.686	0.441	0.310	0.237	0.192	0.4
	0.819	0.772	0.719	0.663	0.608	0.400	0.290	0.226	0.184	0.6
	0.792	0.725	0.658	0.598	0.544	0.364	0.269	0.212	0.175	0.8
	0.764	0.678	0.605	0.544	0.493	0.332	0.250	0.200	0.167	1.0
0.8	0.788	0.788	0.788	0.788	0.788	0.499	0.333	0.250	0.200	0.0
	0.784	0.781	0.774	0.763	0.743	0.480	0.326	0.247	0.198	0.2
	0.776	0.760	0.738	0.708	0.667	0.440	0.310	0.237	0.192	0.4
	0.763	0.730	0.691	0.645	0.597	0.399	0.290	0.226	0.184	0.6
	0.744	0.694	0.639	0.587	0.537	0.363	0.269	0.212	0.175	0.8
	0.723	0.656	0.592	0.536	0.488	0.332	0.250	0.200	0.167	1.0
1.0	0.715	0.715	0.715	0.715	0.715	0.499	0.333	0.250	0.200	0.0
	0.713	0.711	0.708	0.701	0.691	0.479	0.326	0.247	0.198	0.2
	0.708	0.699	0.686	0.667	0.640	0.439	0.310	0.237	0.192	0.4
	0.700	0.680	0.653	0.620	0.581	0.398	0.289	0.226	0.184	0.6
	0.688	0.655	0.615	0.571	0.529	0.362	0.268	0.212	0.175	0.8
	0.674	0.626	0.576	0.527	0.483	0.331	0.250	0.200	0.167	1.0
2.0	0.443	0.443	0.443	0.443	0.443	0.443	0.333	0.250	0.200	0.0
	0.443	0.443	0.443	0.443	0.442	0.433	0.326	0.247	0.198	0.2
	0.443	0.442	0.441	0.440	0.439	0.409	0.308	0.237	0.192	0.4
	0.442	0.441	0.439	0.436	0.433	0.379	0.288	0.225	0.184	0.6
	0.441	0.439	0.435	0.430	0.424	0.350	0.267	0.212	0.175	0.8
	0.440	0.435	0.429	0.422	0.412	0.323	0.248	0.200	0.167	1.0
3.0	0.311	0.311	0.311	0.311	0.311	0.311	0.311	0.250	0.200	0.0
	0.311	0.310	0.310	0.310	0.310	0.310	0.306	0.246	0.198	0.2
	0.310	0.310	0.310	0.310	0.310	0.308	0.294	0.236	0.192	0.4
	0.310	0.310	0.310	0.309	0.309	0.305	0.278	0.224	0.184	0.6
	0.310	0.310	0.309	0.309	0.308	0.298	0.260	0.211	0.175	0.8
	0.310	0.309	0.309	0.308	0.306	0.290	0.243	0.198	0.167	1.0
4.0	0.238	0.238	0.238	0.238	0.238	0.238	0.238	0.238	0.200	0.0
	0.238	0.238	0.238	0.238	0.238	0.237	0.237	0.235	0.197	0.2
	0.238	0.238	0.237	0.237	0.237	0.237	0.236	0.228	0.191	0.4
	0.238	0.237	0.237	0.237	0.237	0.236	0.234	0.218	0.183	0.6
	0.237	0.237	0.237	0.237	0.237	0.236	0.230	0.206	0.174	0.8
	0.237	0.237	0.237	0.237	0.236	0.234	0.225	0.195	0.166	1.0
5.0	0.192	0.192	0.192	0.192	0.192	0.192	0.192	0.192	0.192	0.0
	0.192	0.192	0.192	0.192	0.192	0.192	0.192	0.192	0.191	0.2
	0.192	0.192	0.192	0.192	0.192	0.192	0.192	0.191	0.186	0.4
	0.192	0.192	0.192	0.192	0.192	0.191	0.191	0.190	0.179	0.6
	0.192	0.192	0.192	0.192	0.192	0.191	0.191	0.187	0.171	0.8
	0.192	0.192	0.192	0.192	0.191	0.191	0.189	0.184	0.164	1.0

Table 8-5. Load parameter λ for $\tau_x = 0.5$

$\frac{N_{cr,\varphi}}{N_{cr,y}}$ \ $\frac{N_{cr,\varphi}}{N_{cr,x}}$	0.2	0.4	0.6	0.8	1.0	2.0	3.0	4.0	5.0	τ_y
0.0	1.000	1.000	1.000	1.000	1.000	0.500	0.333	0.250	0.200	-
0.2	0.947	0.947	0.947	0.947	0.947	0.500	0.333	0.250	0.200	0.0
	0.940	0.928	0.909	0.875	0.817	0.481	0.326	0.247	0.198	0.2
	0.919	0.880	0.830	0.770	0.705	0.443	0.310	0.238	0.192	0.4
	0.887	0.819	0.750	0.681	0.619	0.402	0.290	0.226	0.184	0.6
	0.849	0.758	0.678	0.609	0.551	0.365	0.269	0.212	0.175	0.8
	0.809	0.702	0.618	0.552	0.498	0.333	0.250	0.201	0.167	1.0
0.4	0.883	0.883	0.883	0.883	0.883	0.500	0.333	0.250	0.200	0.0
	0.877	0.869	0.856	0.833	0.791	0.481	0.326	0.247	0.198	0.2
	0.861	0.833	0.795	0.747	0.690	0.442	0.310	0.238	0.192	0.4
	0.837	0.785	0.726	0.667	0.610	0.401	0.290	0.226	0.184	0.6
	0.807	0.733	0.663	0.600	0.545	0.364	0.269	0.212	0.175	0.8
	0.775	0.683	0.608	0.545	0.494	0.332	0.250	0.201	0.167	1.0
0.6	0.811	0.811	0.811	0.811	0.811	0.500	0.333	0.250	0.200	0.0
	0.807	0.802	0.794	0.780	0.754	0.480	0.326	0.247	0.198	0.2
	0.796	0.777	0.750	0.716	0.671	0.440	0.310	0.237	0.192	0.4
	0.780	0.742	0.697	0.648	0.598	0.399	0.290	0.226	0.184	0.6
	0.758	0.701	0.643	0.588	0.538	0.363	0.269	0.212	0.175	0.8
	0.734	0.660	0.594	0.537	0.488	0.331	0.250	0.200	0.167	1.0
0.8	0.737	0.737	0.737	0.737	0.737	0.499	0.333	0.250	0.200	0.0
	0.735	0.732	0.727	0.719	0.706	0.479	0.326	0.247	0.198	0.2
	0.729	0.716	0.699	0.676	0.644	0.439	0.310	0.237	0.192	0.4
	0.718	0.692	0.661	0.623	0.582	0.398	0.289	0.226	0.184	0.6
	0.703	0.663	0.618	0.572	0.528	0.361	0.268	0.212	0.175	0.8
	0.686	0.631	0.577	0.526	0.481	0.331	0.250	0.200	0.167	1.0
1.0	0.667	0.667	0.667	0.667	0.667	0.499	0.333	0.250	0.200	0.0
	0.666	0.664	0.661	0.657	0.650	0.477	0.326	0.247	0.198	0.2
	0.662	0.655	0.645	0.630	0.610	0.436	0.309	0.237	0.192	0.4
	0.655	0.639	0.619	0.593	0.562	0.395	0.289	0.226	0.184	0.6
	0.646	0.619	0.588	0.552	0.515	0.360	0.268	0.212	0.175	0.8
	0.635	0.597	0.555	0.513	0.474	0.329	0.250	0.200	0.167	1.0
2.0	0.422	0.422	0.422	0.422	0.422	0.422	0.333	0.250	0.200	0.0
	0.422	0.422	0.422	0.422	0.422	0.415	0.325	0.246	0.198	0.2
	0.422	0.421	0.420	0.419	0.418	0.395	0.307	0.237	0.192	0.4
	0.421	0.420	0.418	0.416	0.412	0.369	0.285	0.225	0.184	0.6
	0.420	0.418	0.414	0.409	0.403	0.342	0.265	0.212	0.175	0.8
	0.419	0.415	0.409	0.402	0.393	0.317	0.247	0.199	0.167	1.0

to be continued

Table 8-5. Continued. Load parameter λ for $\tau_x = 0.5$.

$\frac{N_{cr,\varphi}}{N_{cr,y}}$ \ $\frac{N_{cr,\varphi}}{N_{cr,x}}$	0.2	0.4	0.6	0.8	1.0	2.0	3.0	4.0	5.0	τ_y
	0.301	0.301	0.301	0.301	0.301	0.301	0.301	0.250	0.200	0.0
	0.301	0.301	0.301	0.301	0.301	0.300	0.297	0.246	0.198	0.2
3.0	0.301	0.300	0.300	0.300	0.300	0.298	0.286	0.236	0.191	0.4
	0.300	0.300	0.300	0.300	0.299	0.293	0.271	0.222	0.184	0.6
	0.300	0.300	0.299	0.299	0.298	0.287	0.255	0.209	0.174	0.8
	0.300	0.299	0.298	0.297	0.295	0.280	0.240	0.198	0.166	1.0
	0.232	0.232	0.232	0.232	0.232	0.232	0.232	0.232	0.200	0.0
	0.232	0.232	0.232	0.232	0.232	0.232	0.232	0.229	0.197	0.2
4.0	0.232	0.232	0.232	0.232	0.232	0.231	0.230	0.223	0.191	0.4
	0.232	0.232	0.232	0.232	0.232	0.230	0.228	0.214	0.182	0.6
	0.232	0.232	0.232	0.231	0.231	0.229	0.224	0.203	0.173	0.8
	0.232	0.232	0.231	0.231	0.230	0.227	0.219	0.193	0.165	1.0
	0.188	0.188	0.188	0.188	0.188	0.188	0.188	0.188	0.188	0.0
	0.188	0.188	0.188	0.188	0.188	0.188	0.188	0.188	0.187	0.2
5.0	0.188	0.188	0.188	0.188	0.188	0.188	0.188	0.188	0.183	0.4
	0.188	0.188	0.188	0.188	0.188	0.188	0.188	0.185	0.177	0.6
	0.188	0.188	0.188	0.188	0.188	0.187	0.186	0.183	0.169	0.8
	0.188	0.188	0.188	0.188	0.188	0.187	0.185	0.179	0.162	1.0

Table 8-6. Load parameter λ for $\tau_x = 0.6$.

$\frac{N_{cr,\varphi}}{N_{cr,y}}$ \ $\frac{N_{cr,\varphi}}{N_{cr,x}}$	0.2	0.4	0.6	0.8	1.0	2.0	3.0	4.0	5.0	τ_y
0.0	1.000	1.000	1.000	1.000	1.000	0.500	0.333	0.250	0.200	-
	0.927	0.927	0.927	0.927	0.927	0.500	0.333	0.250	0.200	0.0
	0.920	0.909	0.892	0.862	0.809	0.481	0.326	0.247	0.198	0.2
0.2	0.900	0.864	0.819	0.763	0.700	0.442	0.310	0.238	0.192	0.4
	0.871	0.808	0.742	0.677	0.616	0.401	0.290	0.226	0.184	0.6
	0.836	0.750	0.673	0.606	0.550	0.364	0.269	0.212	0.175	0.8
	0.798	0.695	0.614	0.550	0.496	0.333	0.250	0.200	0.167	1.0
	0.847	0.847	0.847	0.847	0.847	0.500	0.333	0.250	0.200	0.0
	0.842	0.836	0.825	0.806	0.773	0.480	0.326	0.247	0.198	0.2
0.4	0.829	0.805	0.772	0.730	0.680	0.441	0.310	0.238	0.192	0.4
	0.808	0.762	0.711	0.657	0.603	0.400	0.290	0.226	0.184	0.6
	0.782	0.716	0.652	0.593	0.541	0.363	0.269	0.212	0.175	0.8
	0.753	0.671	0.599	0.540	0.490	0.331	0.249	0.200	0.167	1.0

to be continued

Table 8-6. Continued. Load parameter λ for $\tau_x = 0.6$.

$\frac{N_{cr,\varphi}}{N_{cr,y}}$ \ $\frac{N_{cr,\varphi}}{N_{cr,x}}$	0.2	0.4	0.6	0.8	1.0	2.0	3.0	4.0	5.0	τ_y
0.6	0.767	0.767	0.767	0.767	0.767	0.500	0.333	0.250	0.200	0.0
	0.764	0.760	0.753	0.743	0.725	0.479	0.326	0.247	0.198	0.2
	0.756	0.740	0.719	0.691	0.653	0.439	0.310	0.237	0.192	0.4
	0.742	0.711	0.674	0.632	0.587	0.398	0.289	0.226	0.184	0.6
	0.724	0.676	0.626	0.577	0.530	0.361	0.268	0.212	0.175	0.8
	0.704	0.640	0.581	0.529	0.483	0.330	0.249	0.200	0.167	1.0
0.8	0.692	0.692	0.692	0.692	0.692	0.499	0.333	0.250	0.200	0.0
	0.691	0.688	0.684	0.679	0.670	0.477	0.326	0.247	0.198	0.2
	0.685	0.676	0.663	0.645	0.620	0.436	0.309	0.237	0.192	0.4
	0.677	0.656	0.632	0.601	0.567	0.395	0.289	0.226	0.184	0.6
	0.665	0.632	0.595	0.556	0.517	0.360	0.268	0.212	0.175	0.8
	0.651	0.605	0.559	0.515	0.474	0.329	0.249	0.200	0.167	1.0
1.0	0.626	0.626	0.626	0.626	0.626	0.499	0.333	0.250	0.200	0.0
	0.625	0.623	0.621	0.618	0.613	0.475	0.326	0.247	0.198	0.2
	0.621	0.615	0.608	0.597	0.581	0.432	0.309	0.237	0.192	0.4
	0.616	0.603	0.587	0.567	0.542	0.392	0.288	0.226	0.184	0.6
	0.609	0.587	0.561	0.532	0.501	0.357	0.267	0.212	0.175	0.8
	0.601	0.568	0.533	0.498	0.463	0.327	0.249	0.200	0.167	1.0
2.0	0.402	0.402	0.402	0.402	0.402	0.402	0.333	0.250	0.200	0.0
	0.402	0.402	0.402	0.402	0.402	0.396	0.324	0.246	0.198	0.2
	0.402	0.401	0.400	0.399	0.398	0.379	0.305	0.236	0.191	0.4
	0.401	0.400	0.398	0.395	0.392	0.357	0.284	0.224	0.184	0.6
	0.400	0.398	0.394	0.390	0.384	0.333	0.263	0.211	0.175	0.8
	0.398	0.395	0.389	0.383	0.375	0.311	0.245	0.198	0.166	1.0
3.0	0.290	0.290	0.290	0.290	0.290	0.290	0.290	0.250	0.200	0.0
	0.290	0.290	0.290	0.290	0.290	0.289	0.287	0.245	0.197	0.2
	0.290	0.290	0.290	0.290	0.289	0.288	0.278	0.234	0.191	0.4
	0.290	0.290	0.289	0.289	0.288	0.284	0.264	0.221	0.183	0.6
	0.290	0.289	0.288	0.288	0.287	0.278	0.250	0.208	0.174	0.8
	0.289	0.288	0.288	0.286	0.284	0.270	0.236	0.196	0.165	1.0
4.0	0.226	0.226	0.226	0.226	0.226	0.226	0.226	0.226	0.200	0.0
	0.226	0.226	0.226	0.226	0.226	0.226	0.226	0.224	0.197	0.2
	0.226	0.226	0.226	0.226	0.226	0.225	0.224	0.218	0.190	0.4
	0.226	0.226	0.226	0.226	0.226	0.224	0.221	0.209	0.181	0.6
	0.226	0.226	0.226	0.225	0.225	0.222	0.217	0.200	0.172	0.8
	0.226	0.226	0.225	0.225	0.225	0.221	0.212	0.190	0.164	1.0
5.0	0.184	0.184	0.184	0.184	0.184	0.184	0.184	0.184	0.184	0.0
	0.184	0.184	0.184	0.184	0.184	0.184	0.184	0.184	0.183	0.2
	0.184	0.184	0.184	0.184	0.184	0.184	0.184	0.183	0.179	0.4
	0.184	0.184	0.184	0.184	0.184	0.184	0.183	0.181	0.174	0.6
	0.184	0.184	0.184	0.184	0.184	0.184	0.182	0.178	0.167	0.8
	0.184	0.184	0.184	0.184	0.184	0.183	0.181	0.174	0.160	1.0

Table 8-7. Load parameter λ for $\tau_x = 0.7$

$\frac{N_{cr,\varphi}}{N_{cr,y}}$ \ $\frac{N_{cr,\varphi}}{N_{cr,x}}$	0.2	0.4	0.6	0.8	1.0	2.0	3.0	4.0	5.0	τ_y
0.0	1.000	1.000	1.000	1.000	1.000	0.500	0.333	0.250	0.200	-
0.2	0.905	0.905	0.905	0.905	0.905	0.500	0.333	0.250	0.200	0.0
	0.898	0.889	0.874	0.847	0.800	0.481	0.326	0.247	0.198	0.2
	0.881	0.848	0.806	0.753	0.694	0.441	0.310	0.238	0.192	0.4
	0.853	0.795	0.733	0.671	0.612	0.400	0.290	0.226	0.184	0.6
	0.821	0.740	0.667	0.602	0.547	0.364	0.269	0.212	0.175	0.8
	0.786	0.688	0.610	0.547	0.495	0.333	0.250	0.200	0.167	1.0
0.4	0.812	0.812	0.812	0.812	0.812	0.500	0.333	0.250	0.200	0.0
	0.808	0.802	0.793	0.779	0.753	0.480	0.326	0.247	0.198	0.2
	0.796	0.776	0.748	0.712	0.668	0.440	0.310	0.238	0.192	0.4
	0.778	0.739	0.694	0.645	0.595	0.399	0.290	0.226	0.184	0.6
	0.756	0.698	0.640	0.585	0.536	0.362	0.269	0.212	0.175	0.8
	0.731	0.657	0.591	0.534	0.487	0.330	0.249	0.200	0.167	1.0
0.6	0.727	0.727	0.727	0.727	0.727	0.500	0.333	0.250	0.200	0.0
	0.725	0.721	0.716	0.709	0.695	0.478	0.326	0.247	0.198	0.2
	0.718	0.705	0.688	0.665	0.635	0.438	0.310	0.237	0.192	0.4
	0.706	0.681	0.650	0.614	0.574	0.396	0.289	0.226	0.184	0.6
	0.691	0.651	0.608	0.564	0.522	0.360	0.268	0.212	0.175	0.8
	0.674	0.620	0.567	0.519	0.477	0.329	0.249	0.200	0.167	1.0
0.8	0.653	0.653	0.653	0.653	0.653	0.499	0.333	0.250	0.200	0.0
	0.651	0.650	0.646	0.643	0.636	0.476	0.326	0.247	0.198	0.2
	0.647	0.639	0.629	0.615	0.596	0.433	0.309	0.237	0.192	0.4
	0.639	0.623	0.603	0.578	0.550	0.393	0.288	0.226	0.184	0.6
	0.630	0.603	0.573	0.539	0.505	0.357	0.267	0.212	0.175	0.8
	0.619	0.581	0.541	0.502	0.465	0.328	0.249	0.200	0.167	1.0
1.0	0.589	0.589	0.589	0.589	0.589	0.499	0.333	0.250	0.200	0.0
	0.588	0.587	0.585	0.583	0.579	0.472	0.326	0.247	0.198	0.2
	0.585	0.581	0.574	0.566	0.554	0.429	0.308	0.237	0.192	0.4
	0.581	0.570	0.557	0.541	0.521	0.389	0.288	0.225	0.184	0.6
	0.574	0.557	0.536	0.512	0.485	0.354	0.267	0.212	0.175	0.8
	0.567	0.542	0.512	0.482	0.451	0.325	0.248	0.199	0.167	1.0
2.0	0.383	0.383	0.383	0.383	0.383	0.383	0.333	0.250	0.200	0.0
	0.383	0.383	0.383	0.382	0.382	0.378	0.322	0.246	0.198	0.2
	0.383	0.382	0.381	0.381	0.379	0.364	0.302	0.236	0.191	0.4
	0.382	0.381	0.379	0.377	0.374	0.345	0.281	0.223	0.184	0.6
	0.381	0.379	0.375	0.371	0.367	0.324	0.260	0.210	0.174	0.8
	0.380	0.376	0.371	0.365	0.359	0.304	0.243	0.198	0.166	1.0

to be continued

Table 8-7. Continued. Load parameter λ for $\tau_x = 0.7$.

$N_{cr,\varphi}/N_{cr,y}$ \ $N_{cr,\varphi}/N_{cr,x}$	0.2	0.4	0.6	0.8	1.0	2.0	3.0	4.0	5.0	τ_y
3.0	0.280	0.280	0.280	0.280	0.280	0.280	0.280	0.250	0.200	0.0
	0.280	0.280	0.280	0.280	0.280	0.279	0.277	0.244	0.197	0.2
	0.280	0.279	0.279	0.279	0.279	0.277	0.269	0.233	0.191	0.4
	0.279	0.279	0.278	0.278	0.278	0.274	0.257	0.219	0.183	0.6
	0.279	0.278	0.278	0.277	0.276	0.268	0.244	0.206	0.174	0.8
	0.279	0.278	0.277	0.276	0.274	0.260	0.231	0.195	0.165	1.0
4.0	0.219	0.219	0.219	0.219	0.219	0.219	0.219	0.219	0.200	0.0
	0.219	0.219	0.219	0.219	0.219	0.219	0.219	0.218	0.196	0.2
	0.219	0.219	0.219	0.219	0.219	0.219	0.217	0.212	0.188	0.4
	0.219	0.219	0.219	0.219	0.219	0.218	0.215	0.205	0.180	0.6
	0.219	0.219	0.219	0.219	0.218	0.216	0.211	0.195	0.171	0.8
	0.219	0.219	0.219	0.218	0.218	0.214	0.206	0.187	0.163	1.0
5.0	0.180	0.180	0.180	0.180	0.180	0.180	0.180	0.180	0.180	0.0
	0.180	0.180	0.180	0.180	0.180	0.180	0.180	0.180	0.179	0.2
	0.180	0.180	0.180	0.180	0.180	0.180	0.180	0.178	0.175	0.4
	0.180	0.180	0.180	0.180	0.180	0.180	0.178	0.177	0.170	0.6
	0.180	0.180	0.180	0.180	0.180	0.180	0.178	0.174	0.164	0.8
	0.180	0.180	0.180	0.180	0.180	0.179	0.176	0.171	0.157	1.0

Table 8-8. Load parameter λ for $\tau_x = 0.8$.

$N_{cr,\varphi}/N_{cr,y}$ \ $N_{cr,\varphi}/N_{cr,x}$	0.2	0.4	0.6	0.8	1.0	2.0	3.0	4.0	5.0	τ_y
0.0	1.000	1.000	1.000	1.000	1.000	0.500	0.333	0.250	0.200	-
0.2	0.882	0.882	0.882	0.882	0.882	0.500	0.333	0.250	0.200	0.0
	0.876	0.868	0.854	0.831	0.789	0.481	0.326	0.247	0.198	0.2
	0.860	0.831	0.792	0.744	0.688	0.441	0.310	0.237	0.192	0.4
	0.836	0.782	0.724	0.665	0.609	0.400	0.290	0.226	0.184	0.6
	0.805	0.730	0.660	0.598	0.544	0.364	0.269	0.212	0.175	0.8
	0.773	0.681	0.605	0.543	0.492	0.332	0.250	0.200	0.167	1.0
0.4	0.778	0.778	0.778	0.778	0.778	0.500	0.333	0.250	0.200	0.0
	0.774	0.770	0.763	0.751	0.730	0.479	0.326	0.247	0.198	0.2
	0.765	0.747	0.725	0.694	0.655	0.439	0.310	0.237	0.192	0.4
	0.750	0.716	0.676	0.633	0.587	0.398	0.289	0.226	0.184	0.6
	0.730	0.679	0.626	0.576	0.529	0.361	0.268	0.212	0.175	0.8
	0.709	0.641	0.581	0.528	0.482	0.329	0.249	0.200	0.167	1.0

to be continued

Table 8-8. Continued. Load parameter λ for $\tau_x = 0.8$.

$\frac{N_{cr,\varphi}}{N_{cr,y}}$ \ $\frac{N_{cr,\varphi}}{N_{cr,x}}$	0.2	0.4	0.6	0.8	1.0	2.0	3.0	4.0	5.0	τ_y
0.6	0.690	0.690	0.690	0.690	0.690	0.500	0.333	0.250	0.200	0.0
	0.688	0.685	0.681	0.646	0.666	0.477	0.326	0.246	0.198	0.2
	0.682	0.672	0.658	0.640	0.615	0.435	0.309	0.237	0.192	0.4
	0.673	0.652	0.626	0.595	0.561	0.394	0.288	0.226	0.184	0.6
	0.660	0.626	0.589	0.550	0.512	0.358	0.267	0.212	0.175	0.8
	0.647	0.600	0.553	0.510	0.470	0.327	0.248	0.199	0.167	1.0
0.8	0.617	0.617	0.617	0.617	0.617	0.499	0.333	0.250	0.200	0.0
	0.616	0.614	0.612	0.609	0.604	0.474	0.326	0.246	0.198	0.2
	0.612	0.606	0.598	0.587	0.571	0.430	0.309	0.237	0.192	0.4
	0.606	0.593	0.577	0.556	0.532	0.390	0.288	0.225	0.184	0.6
	0.598	0.576	0.550	0.522	0.492	0.355	0.267	0.212	0.175	0.8
	0.588	0.557	0.523	0.489	0.456	0.325	0.248	0.199	0.167	1.0
1.0	0.556	0.556	0.556	0.556	0.556	0.499	0.333	0.250	0.200	0.0
	0.555	0.554	0.553	0.551	0.549	0.467	0.326	0.246	0.198	0.2
	0.553	0.549	0.544	0.537	0.529	0.424	0.308	0.237	0.192	0.4
	0.550	0.541	0.530	0.517	0.501	0.384	0.287	0.225	0.184	0.6
	0.544	0.529	0.512	0.492	0.470	0.351	0.266	0.212	0.175	0.8
	0.537	0.517	0.492	0.467	0.440	0.322	0.247	0.199	0.167	1.0
2.0	0.365	0.365	0.365	0.365	0.365	0.365	0.333	0.250	0.200	0.0
	0.365	0.365	0.365	0.364	0.364	0.361	0.320	0.246	0.198	0.2
	0.365	0.364	0.364	0.363	0.362	0.350	0.298	0.236	0.191	0.4
	0.364	0.363	0.361	0.360	0.357	0.333	0.278	0.222	0.184	0.6
	0.364	0.361	0.358	0.355	0.351	0.315	0.258	0.209	0.174	0.8
	0.362	0.358	0.354	0.349	0.343	0.297	0.240	0.197	0.166	1.0
3.0	0.270	0.270	0.270	0.270	0.270	0.270	0.270	0.250	0.200	0.0
	0.269	0.269	0.269	0.269	0.269	0.269	0.267	0.243	0.197	0.2
	0.269	0.269	0.269	0.269	0.268	0.267	0.260	0.230	0.191	0.4
	0.269	0.269	0.268	0.268	0.267	0.263	0.250	0.217	0.182	0.6
	0.269	0.268	0.267	0.267	0.266	0.258	0.238	0.204	0.173	0.8
	0.268	0.267	0.267	0.265	0.264	0.251	0.226	0.193	0.164	1.0
4.0	0.212	0.212	0.212	0.212	0.212	0.212	0.212	0.212	0.200	0.0
	0.212	0.212	0.212	0.212	0.212	0.212	0.212	0.211	0.195	0.2
	0.212	0.212	0.212	0.212	0.212	0.212	0.211	0.206	0.187	0.4
	0.212	0.212	0.212	0.212	0.212	0.211	0.208	0.200	0.178	0.6
	0.212	0.212	0.212	0.212	0.212	0.209	0.204	0.191	0.169	0.8
	0.212	0.212	0.212	0.212	0.211	0.207	0.200	0.184	0.161	1.0
5.0	0.176	0.176	0.176	0.176	0.176	0.176	0.176	0.176	0.176	0.0
	0.176	0.176	0.175	0.175	0.175	0.175	0.175	0.175	0.174	0.2
	0.175	0.175	0.175	0.175	0.175	0.175	0.175	0.174	0.171	0.4
	0.175	0.175	0.175	0.175	0.175	0.175	0.174	0.172	0.167	0.6
	0.175	0.175	0.175	0.175	0.175	0.174	0.173	0.169	0.160	0.8
	0.175	0.175	0.175	0.175	0.174	0.174	0.171	0.166	0.154	1.0

Table 8-9. Load parameter λ for $\tau_x = 0.9$

$\frac{N_{cr,\varphi}}{N_{cr,x}}$ / $\frac{N_{cr,\varphi}}{N_{cr,y}}$	0.2	0.4	0.6	0.8	1.0	2.0	3.0	4.0	5.0	τ_y
0.0	1.000	1.000	1.000	1.000	1.000	0.500	0.333	0.250	0.200	-
0.2	0.859	0.859	0.859	0.859	0.859	0.500	0.333	0.250	0.200	0.0
	0.853	0.846	0.834	0.814	0.778	0.480	0.326	0.247	0.198	0.2
	0.839	0.812	0.778	0.733	0.681	0.440	0.310	0.237	0.192	0.4
	0.816	0.767	0.714	0.658	0.604	0.399	0.290	0.226	0.184	0.6
	0.788	0.719	0.653	0.594	0.541	0.363	0.269	0.212	0.175	0.8
	0.759	0.672	0.600	0.540	0.490	0.332	0.250	0.200	0.167	1.0
0.4	0.746	0.746	0.746	0.746	0.746	0.500	0.333	0.250	0.200	0.0
	0.743	0.739	0.733	0.724	0.708	0.478	0.326	0.247	0.198	0.2
	0.735	0.720	0.701	0.674	0.641	0.437	0.309	0.237	0.192	0.4
	0.722	0.692	0.658	0.619	0.578	0.396	0.288	0.226	0.184	0.6
	0.705	0.660	0.613	0.567	0.523	0.359	0.267	0.212	0.175	0.8
	0.685	0.626	0.571	0.521	0.478	0.328	0.249	0.199	0.167	1.0
0.6	0.657	0.657	0.657	0.657	0.657	0.500	0.333	0.250	0.200	0.0
	0.655	0.653	0.650	0.645	0.638	0.475	0.326	0.246	0.198	0.2
	0.650	0.642	0.630	0.616	0.595	0.431	0.309	0.237	0.192	0.4
	0.643	0.625	0.603	0.577	0.547	0.391	0.288	0.226	0.184	0.6
	0.632	0.603	0.571	0.537	0.502	0.355	0.267	0.212	0.175	0.8
	0.619	0.580	0.539	0.499	0.463	0.325	0.247	0.199	0.167	1.0
0.8	0.585	0.585	0.585	0.585	0.585	0.499	0.333	0.250	0.200	0.0
	0.584	0.583	0.581	0.578	0.574	0.470	0.326	0.246	0.198	0.2
	0.581	0.576	0.569	0.560	0.548	0.426	0.308	0.237	0.192	0.4
	0.576	0.565	0.551	0.535	0.515	0.386	0.287	0.225	0.184	0.6
	0.570	0.551	0.529	0.505	0.479	0.352	0.266	0.212	0.175	0.8
	0.561	0.535	0.506	0.476	0.446	0.323	0.247	0.199	0.167	1.0
1.0	0.527	0.527	0.527	0.527	0.527	0.499	0.333	0.250	0.200	0.0
	0.526	0.525	0.524	0.522	0.521	0.461	0.325	0.246	0.198	0.2
	0.524	0.521	0.517	0.512	0.504	0.418	0.307	0.236	0.192	0.4
	0.521	0.514	0.505	0.495	0.481	0.380	0.286	0.225	0.184	0.6
	0.516	0.505	0.490	0.473	0.454	0.347	0.265	0.211	0.175	0.8
	0.511	0.494	0.473	0.451	0.427	0.319	0.246	0.199	0.167	1.0
2.0	0.349	0.349	0.349	0.349	0.349	0.349	0.333	0.250	0.200	0.0
	0.349	0.349	0.348	0.348	0.348	0.345	0.316	0.245	0.198	0.2
	0.348	0.348	0.347	0.347	0.346	0.336	0.295	0.235	0.191	0.4
	0.347	0.346	0.345	0.343	0.341	0.322	0.274	0.222	0.183	0.6
	0.347	0.344	0.342	0.340	0.336	0.305	0.255	0.209	0.174	0.8
	0.346	0.343	0.339	0.334	0.329	0.289	0.238	0.196	0.165	1.0

to be continued

Table 8-9. Continued. Load parameter λ for $\tau_x = 0.9$.

$\frac{N_{cr,\varphi}}{N_{cr,y}}$ \ $\frac{N_{cr,\varphi}}{N_{cr,x}}$	0.2	0.4	0.6	0.8	1.0	2.0	3.0	4.0	5.0	τ_y
3.0	0.260	0.260	0.260	0.260	0.260	0.260	0.260	0.250	0.200	0.0
	0.260	0.260	0.259	0.259	0.259	0.259	0.257	0.240	0.197	0.2
	0.259	0.259	0.259	0.259	0.258	0.258	0.251	0.227	0.190	0.4
	0.259	0.259	0.258	0.258	0.257	0.255	0.243	0.215	0.181	0.6
	0.259	0.258	0.257	0.257	0.256	0.251	0.232	0.202	0.172	0.8
	0.259	0.257	0.257	0.256	0.253	0.244	0.221	0.191	0.164	1.0
4.0	0.206	0.206	0.206	0.206	0.206	0.206	0.206	0.206	0.200	0.0
	0.206	0.206	0.206	0.206	0.206	0.206	0.205	0.205	0.194	0.2
	0.206	0.206	0.206	0.206	0.206	0.205	0.204	0.201	0.185	0.4
	0.206	0.206	0.205	0.205	0.205	0.204	0.202	0.195	0.176	0.6
	0.206	0.205	0.205	0.205	0.205	0.202	0.198	0.187	0.167	0.8
	0.206	0.205	0.205	0.205	0.205	0.201	0.194	0.180	0.160	1.0
5.0	0.171	0.171	0.171	0.171	0.171	0.171	0.171	0.171	0.171	0.0
	0.171	0.171	0.171	0.171	0.171	0.171	0.171	0.171	0.170	0.2
	0.171	0.171	0.171	0.171	0.171	0.171	0.171	0.170	0.167	0.4
	0.171	0.171	0.171	0.171	0.171	0.170	0.169	0.167	0.163	0.6
	0.171	0.171	0.171	0.171	0.171	0.170	0.168	0.164	0.157	0.8
	0.171	0.171	0.171	0.171	0.170	0.169	0.167	0.162	0.152	1.0

Table 8-10. Load parameter λ for $\tau_x = 1.0$.

$\frac{N_{cr,\varphi}}{N_{cr,y}}$ \ $\frac{N_{cr,\varphi}}{N_{cr,x}}$	0.2	0.4	0.6	0.8	1.0	2.0	3.0	4.0	5.0	τ_y
0.0	1.000	1.000	1.000	1.000	1.000	0.500	0.333	0.250	0.200	-
0.2	0.836	0.836	0.836	0.836	0.836	0.500	0.333	0.250	0.200	0.0
	0.831	0.824	0.814	0.796	0.765	0.480	0.326	0.247	0.198	0.2
	0.818	0.794	0.763	0.722	0.674	0.440	0.310	0.237	0.192	0.4
	0.797	0.753	0.703	0.650	0.599	0.398	0.289	0.226	0.184	0.6
	0.772	0.708	0.646	0.588	0.537	0.362	0.268	0.212	0.175	0.8
	0.743	0.663	0.594	0.536	0.487	0.331	0.249	0.199	0.166	1.0
0.4	0.716	0.716	0.716	0.716	0.716	0.500	0.333	0.250	0.200	0.0
	0.713	0.710	0.705	0.698	0.685	0.478	0.326	0.247	0.198	0.2
	0.706	0.694	0.677	0.655	0.626	0.435	0.309	0.237	0.192	0.4
	0.695	0.670	0.640	0.605	0.567	0.395	0.288	0.226	0.184	0.6
	0.680	0.640	0.599	0.557	0.516	0.358	0.267	0.212	0.175	0.8
	0.663	0.609	0.559	0.513	0.471	0.327	0.248	0.199	0.166	1.0

to be continued

Table 8-10. Continued. Load parameter λ for $\tau_x = 1.0$.

$N_{cr,\varphi}/N_{cr,y}$ \ $N_{cr,\varphi}/N_{cr,x}$	0.2	0.4	0.6	0.8	1.0	2.0	3.0	4.0	5.0	τ_y
	0.626	0.626	0.626	0.626	0.626	0.500	0.333	0.250	0.200	0.0
	0.625	0.622	0.620	0.616	0.611	0.473	0.326	0.246	0.198	0.2
0.6	0.620	0.613	0.604	0.591	0.575	0.429	0.309	0.237	0.192	0.4
	0.614	0.598	0.581	0.558	0.533	0.389	0.288	0.226	0.184	0.6
	0.605	0.581	0.553	0.522	0.491	0.354	0.267	0.212	0.175	0.8
	0.594	0.559	0.524	0.488	0.454	0.324	0.247	0.198	0.166	1.0
	0.557	0.557	0.557	0.557	0.557	0.499	0.333	0.250	0.200	0.0
	0.555	0.554	0.553	0.551	0.548	0.466	0.326	0.246	0.198	0.2
0.8	0.553	0.549	0.543	0.536	0.526	0.422	0.308	0.237	0.192	0.4
	0.549	0.539	0.528	0.514	0.497	0.382	0.286	0.225	0.184	0.6
	0.543	0.527	0.509	0.488	0.466	0.349	0.265	0.212	0.175	0.8
	0.536	0.513	0.488	0.461	0.435	0.320	0.246	0.198	0.166	1.0
	0.500	0.500	0.500	0.500	0.500	0.499	0.333	0.250	0.200	0.0
	0.500	0.499	0.498	0.497	0.495	0.453	0.325	0.246	0.198	0.2
1.0	0.498	0.495	0.492	0.488	0.482	0.411	0.306	0.236	0.191	0.4
	0.495	0.489	0.482	0.473	0.462	0.374	0.284	0.224	0.184	0.6
	0.491	0.481	0.469	0.455	0.439	0.343	0.263	0.211	0.174	0.8
	0.487	0.471	0.454	0.435	0.415	0.315	0.244	0.198	0.165	1.0
	0.333	0.333	0.333	0.333	0.333	0.333	0.333	0.250	0.200	0.0
	0.333	0.333	0.333	0.333	0.333	0.330	0.311	0.245	0.197	0.2
2.0	0.333	0.333	0.332	0.331	0.330	0.322	0.290	0.234	0.191	0.4
	0.333	0.331	0.330	0.329	0.326	0.310	0.270	0.221	0.183	0.6
	0.332	0.329	0.327	0.325	0.322	0.296	0.251	0.207	0.174	0.8
	0.331	0.327	0.324	0.320	0.316	0.281	0.234	0.195	0.164	1.0
	0.250	0.250	0.250	0.250	0.250	0.250	0.250	0.250	0.200	0.0
	0.250	0.250	0.250	0.250	0.250	0.250	0.248	0.237	0.197	0.2
3.0	0.250	0.250	0.250	0.250	0.249	0.248	0.243	0.224	0.189	0.4
	0.250	0.249	0.249	0.249	0.248	0.245	0.235	0.212	0.181	0.6
	0.250	0.249	0.248	0.248	0.247	0.240	0.226	0.199	0.171	0.8
	0.249	0.248	0.247	0.246	0.244	0.234	0.215	0.188	0.161	1.0
	0.200	0.200	0.200	0.200	0.200	0.200	0.200	0.200	0.200	0.0
	0.200	0.200	0.200	0.200	0.199	0.199	0.199	0.198	0.191	0.2
4.0	0.200	0.199	0.199	0.199	0.199	0.199	0.198	0.195	0.183	0.4
	0.199	0.199	0.199	0.199	0.199	0.198	0.195	0.189	0.174	0.6
	0.199	0.199	0.199	0.198	0.198	0.196	0.192	0.183	0.165	0.8
	0.199	0.199	0.198	0.198	0.198	0.195	0.188	0.175	0.157	1.0
	0.167	0.167	0.167	0.167	0.167	0.167	0.167	0.167	0.167	0.0
	0.167	0.167	0.167	0.167	0.167	0.167	0.166	0.166	0.165	0.2
5.0	0.167	0.167	0.167	0.167	0.166	0.166	0.166	0.165	0.163	0.4
	0.167	0.167	0.167	0.167	0.166	0.166	0.164	0.163	0.159	0.6
	0.166	0.166	0.166	0.166	0.166	0.166	0.164	0.160	0.153	0.8
	0.166	0.166	0.166	0.166	0.165	0.165	0.161	0.157	0.148	1.0

8.4 MATHEMATICAL MODELLING OF TEST DATA

8.4.1 Preamble

This article may seem a little unusual in a book on structural design. However, it seems justified on the grounds that tools which are not commonplace in the design office are required to solve some aspects of the problem of stability in large buildings. Therefore it is proper that the rationale for proposals to improve this situation should be available. A sound understanding of the principles involved in any method for structural design is essential to ensure its proper use. This being so, no further apology is made.

8.4.2 Introduction

In any discussion concerning the properties of mechanical systems, it is reasonable to suppose that mechanical elements exist in environments which are hierarchically structured. For this argument to be sustained, it is implicit that such elements have also to comprise a similar internal structure. Such a premise is of course endemic to the reductionist foundations of analytical science. It is also appropriate to call for the interpretation of natural phenomena in terms of purely mechanistic behaviour. For a theory of mechanical behaviour to be of any value to a designer of structural systems, it must be possible to discuss the expression of causal relationships mathematically and to produce laws, that is mathematical models, in a conventional form such as Hooke's Law. It is reasonable also to assume that continuous laws should operate between hierarchical boundaries or more tautologically, continuous relationships in behaviour which can be modelled mathematically, will exist between discontinuities.

It is proposed to discuss briefly the form of such laws here and to demonstrate that a consistent approach to their construction must lead to consistency of form.

8.4.3 The Elements of Law

For the purposes of this discussion, the word 'law' is defined as a mathematical model of a continuous mechanical causal relationship. In setting up a such law to represent cause and effect relationships, it is generally required to make a statement of the form:

$$F(X) = F(Y)$$

for

$$x_1 < X < x_2 \quad \text{and} \quad y_1 < Y < y_2$$

and where X and Y could be expressed thus:-

$$X = f(a,b,c,...) \text{ and } Y = f(A,B,C,...)$$

In particular instances, the following two examples have been obtained and can be used as illustrations:-

1. Hooke's Law

$$\sigma = Ee, \tag{8.52}$$

where neither the yield stress (σ) or strain (e) are exceeded.

2. Plate Equilibrium Equations

$$\frac{2M_x}{x^2} + \frac{2M_y}{y^2} \pm \frac{2M_{xy}}{x\,y} = -P(x,y),$$

which holds provided the stated boundary conditions for M_x, M_y and M_{xy} can be satisfied at the edges of the plate and provided that $P(x,y)$ is physically obtainable.

Now the form of the expression (8.52) i.e. one parameter equals another parameter, can be justified on the same basis as that on which the hierarchical structure was adopted namely that any two entities which are to exist in the world, can be related to each other in at least one way. For the relationship of equality to be proven (established) or demonstrated in a particular instance, $F(X)$ and $F(Y)$ must be expressed in the same terms. This is precisely paralleled by the phenomenon in the physical world of interactions between objects only taking place at a common element level. Such a characteristic has been created in advanced societies through the device of money for the evaluation and thus exchange of apparently disparate entities such as food and labour. In most mechanical interactions it is neither necessary nor indeed possible to identify the common element involved in the information exchange between the two parts of the causal system. It is the bulk properties of the parts which are of interest e.g. the gas pressure and the balloon strength not the individual gas molecules and the molecules of the rubber - if indeed that is the level at which interaction takes place!

The common currencies in which causes and effects must be expressed, fall, in the first place, into three categories.

1. Physical
2. Numerical
3. Conceptual

By way of illustrating these categories, consider Pythagoras' Theorem concerning the sides of a right-angled triangle. This can be proven in many ways. Firstly consider a physical proof, a right-angled triangle is drawn on a piece of paper and a square constructed on each of the sides, then, the two small squares are cut out and they can, with further cutting, be fitted exactly into the largest square. In this proof, the statements in the theorem are given a common material substance and can therefore be directly related to each other. For a proof in the second category, the elements of the theorem must be translated into number. This must be done by choosing a unit by type (not size) that is a length unit or an area unit etc., then an operational technique for measurement and finally a scale of measurement for *each* statement. By these means it is possible to obtain a number representing the size of a square constructed on each side of a particular triangle. For the moment, it is sufficient to define scale as a rule for making numerical assignments [Ellis, 1966]. The theorem can now be written down in numbers with the appropriate operational signs ie plus, minus etc., between them, and with constants to ensure that the equality is true. The sources of these constants are the choice of units and the choice of scales. The three choices listed for numerical proofs will be discussed in detail later.

$$L_1/L_3 = \sin\Theta \qquad L_2/L_3 = \cos\Theta$$

$$\sin^2\Theta + \cos^2\Theta = 1$$

$$\text{then } L_1^2 + L_2^2 = L_3^2$$

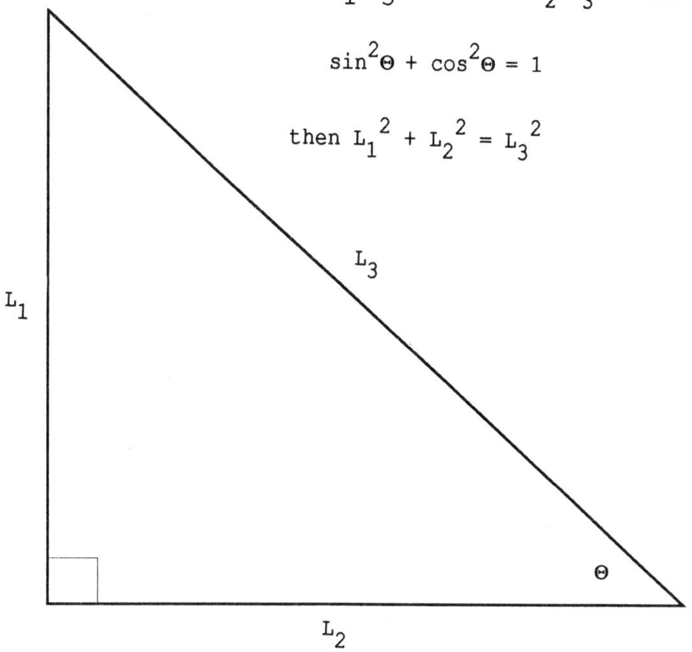

Fig. 8-4. Demonstration of conceptual theory proof

Now for a proof in the conceptual category, consider the triangle illustrated in Fig. 8-4. The sides *L1*, *L2* and *L3* can be put in terms of the angle - and by these means, as shown beside the figure, the theorem can be proven or at least demonstrated. The real proof however has been pushed to a lower level, that is, this proof rests on existing proofs of the laws of trigonometry used. These can only be provided by separate proofs in either the numerical or physical categories which have been outlined already. Conceptual proofs are, of course, the most powerful type available but it must be realised that their justification rests on proofs in the other categories. In complex arguments, all these proofs cannot be gone through on every occasion so a certain range of proofs have to be accepted, as understood, and justified by formal logic chains.

Of the three categories of proof which have been discussed the physical category is the most limited in potential application, since in many situations it will not be possible to make the physical comparison required to perform a proof of this type.

Consider for example, all problems where time is involved, entities which do not exist at the same time cannot be physically compared. If the strengths of concrete cubes made from the same mix and tested at different ages are to be compared, the strength of each cube can only be related to the strengths of the others by translating them into a parameter which is assumed to be independent of time, i.e. constant. In this way, the properties of one cube can be compared with those of another, but a physical comparison could never be made since the cubes cannot exist in the state which is of concern, at identical times - by definition in this case. By the same reasoning, neither can two things be put in an identical place for comparison.

This leaves the two remaining categories that is numerical and conceptual, of these the numerical is the most important for experimental science and since this is the main concern of this note the discussion will concentrate on this in greater detail.

8.4.4 Numerical Proofs

8.4.4.1 Units

For the purposes of this argument, it is accepted that properties of entities can be described quantitatively. Ellis [1966] has discussed the concept of quantity at some length in his book "Basic Concepts of Measurement." Quantitative descriptions of entities are binary in character consisting of a number and a unit name. The unit name relates the qualitative description to the property concerned, that is length, resistance etc., and usually implies that a unit of a particular size has been chosen such as an inch or ohm. In some cases, as for example in swimming, the unit name and property name are the same, i.e. swimming so many lengths of a pool. A unit is then a standard in which a quantity is expressed and as such is **indivisible**. It follows therefore that contrary to the common assumption, there is not the option of choice in size of unit, since in any particular instance, the property which is being measured may not be expressible in an integral number of arbitrarily chosen units. This arbitrariness is in relation to the object of observation but not, of course, to the observer. To overcome this problem in practice, chosen units are divided into 1/10ths, 1/100ths, etc and so on. Therefore the size of the **real** (actual) measuring unit used is determined firstly by the size of what is required to be measured, that is whether it can be described in terms of integral numbers of units and secondly by the level of discrimination [Ellis, 1966] in the measuring operation, that is the smallest unit which is distinguishable by the measuring devices as used by the observer. Now the relationship between the assumed unit and the real unit is that the former is a whole number of the latter or alternatively if x is the real unit, nx is the chosen unit, where n is an integral number. There are no constraints which force the use of this relationship, but it is obviously one which is practical and simple to use.

The numerical part of the quantitative description of a property represents the number of units of the property which the entity being observed is considered to possess.

In the case of some properties, the zero from which counting the number of units should be started is self-evident. Consider for example length, if the length of a box is to be measured then the number of length units involved is quite clear.

If, however, temperature is to be measured it is not quite so obvious where the zero should be taken. As is well known, the freezing point of water has been chosen as the zero for the Centigrade temperature scale.

It was not until the nature of temperature was more fully understood that the existence of a more significant zero at approximately -273°C was appreciated. This means that temperatures measured on the Centigrade scale do not indicate the real number of temperature units involved but about 273 units more than this.

Measurement on the Centigrade scale is thus equivalent to measurement with a foot ruler where instead of the markings being 0, 1, 2, 3, 4, from one end in the usual way, what is normally called 3 inches would be 0 and 4-1, 5-2 and so on, so that the measurements made with this ruler would always be 3 inches in error.

In the case of temperature measurement with an incorrect zero, it is obviously possible to work within limited terms of reference but it has to be remembered that the true numbers of units involved are not being measured (counted) and therefore only changes or differences in temperature can be sensibly discussed. In which instances, the zero errors are cancelled on subtraction. If an arbitrary zero for one measurement is chosen when relating two parallel properties of an object eg. strength and density, then the apparent relationship between the unit of measurement of one property and that of the other will almost invariably be wrong, but dependent upon the off-chance that the arbitrary choice has fortuitously fallen on the proper zero.

Consider the following problems where two properties have been measured in a units and b units respectively. In one particular case, it may be found that:

$$30a = 90b$$

If this is to be a non-trivial law (i.e. true at more than one point) it should be possible to deduce the relation between the two units by simple division:-

$$a = 3b$$

and thus predict a new state at, for example $37a$.

Now using the unit relationship, this implies a new state:-

$$37a = 111b$$

If however $90b$ was measured from a false origin at $+60b$ then the real relationship between a and b is:-

$$1 \text{ x } a = 5b$$

thus

$$37a = 185b$$

and if the original error of $60b$ is now subtracted to get back to the arbitrary origin, what should be found is that:-

$$37a = 125b$$

Now the only way by which this could have been arrived at in the beginning, would have been to have written the relationship between a and b from the first measurement as:

$$1 \times a = 3b + 2b$$

or more properly

$$1 \times a = 3(1 \times b + 2 \times b/3)$$

That is by providing an automatic correction for the zero error in the form of an extra constant.

In this discussion of zero error, a major source of complication in the preparation of a law has been identified. Whenever the number of units involved in a measurement is miscounted, an arbitrary constant is introduced into any consequent equations which might be developed, unless these are in terms of differences. Since many problems involve several measurements on different scales with complex units, if there are zero errors in each of these scales, an equivalent number of arbitrary constants is automatically introduced and it is not difficult to see how laws relating numbers of complex units which have zero errors will inevitably be complicated.

8.4.4.2 Operational Techniques for Measurement

The act of measurement of a property requires some sort of physical operation to take place. The techniques involved can be divided broadly into two sections - direct and indirect.

The use of a conventional ruler to measure length is an example of direct measurement where there is a series of standard units marked out on a rod which can then be physically placed in contact with whatever object is to be measured. By this means, the number of units required in the length measured can be counted. This technique of direct measurement of length is only applicable to a narrow range of length. It is inappropriate or more usually impossible for what are, in human terms, very short or very long lengths. For example, in the case of very long lengths as might be encountered in land surveying, tellurometry is often used. In this technique, a radio wave is reflected along the line to be measured and the time for its journey measured and since its velocity is known, the path length can be deduced. Now this is an indirect means of measurement which requires the calibration of the measuring instrument in terms of length units even though it will be measuring some other characteristic of the space between the two ends of the line (i.e. the time taken for a pulse of radio waves to travel twice the length to be measured). The calibration of the measuring instrument can be undertaken either directly by trial or known length or by using the knowledge of the mechanism by which the instrument operates and deducing the relationship between what it measures directly and length. This latter course undoubtedly requires measurements of other properties such as wave length or time etc. If any of these measurements are not started at real zeros then unknown constants will be automatically introduced. Indirect measurement will, in general, be related with complex units which are a common source of problems in the development of simple laws since it is difficult to understand their physical associations. These complex units will be discussed in detail later.

8.4.4.3 Scale of Measurement

A scale was defined above as a rule for making numerical assignments. This definition is however, too general to be very useful. The precise conditions which must be satisfied for the construction of scales of measurement, have to be identified. The rules used must be what Ellis calls non-degenerate meaning that rules such as "assign the number two to everything" must not be invoked, since they will not convey any useful information. Also the rules must ensure that the same numbers are assigned to the same properties in similar situations. That is, the use of the same ruler on different occasions on the same length, will result in the same numerical result.

It appears that these rules are automatically satisfied if the relationship, which Ellis claims is necessary for the existence of quantity is accepted, namely, that one can establish a set of linear ordering relationships between measured elements. This set of relationships follows from a simple quantitative assessment of a particular situation.

It seems impossible to establish such a set if a unit or standard and secondly a means for counting the number of these standards which are involved in a measurement either explicitly as is done in length or implicitly as in colour, have not firstly been set up. There is a pleasing correspondence here between the establishment of this set of linear ordering relationships and the requirement for a hierarchical structure as described above, that is that for two entities to exist in a hierarchical structure, it must be possible to place them in ranking order by virtue of some common property. It is essential that these parallel features in the physical model and the mathematical model should be detectable, since as soon as this correspondence is lost the real significance of the mathematical law will be impossible to appreciate.

8.4.4.4 The Problem of Dimensions

This note is concerned with causal relationships. In effect, this means that measurements made on one scale need to be related to those made on another and the significance of any correspondence which it is required to model, must be established.

Returning to the example given in the section on units, this model is quite understandable if the units a and b are, for example, both length units. In such a situation, one unit (the smallest) can be used as a measure of the larger one. This gives the statement $30a = 90b$ have some physical and demonstrable meaning, since dimensionally identical (eg length) units are being used on both sides of the equation. If, however, one was a length unit and the other a volume unit the problem is somewhat more complicated since it is not obvious how one can compare a length with a volume in this way. The realms of dimensional analysis are now being entered and what has to be done is to establish how comparability can be obtained in the formulation of a mathematical law. Ideally, this requires a physical understanding of the relationship between one property of an object and another. In mathematical terms, equations are required to be made homogeneous. There appear to be three ways in which this homogeneity can be obtained. Firstly, by the use of empirical constants such as Young's Modulus which themselves have dimensions. Secondly, by a fundamental understanding of the relationship between the measured properties such as a volume which is length cubed (it could also be related to the weight of a specified material), although it could be argued that in this case it is simply a matter of definition, i.e. that volume has been defined as length cubed. Thirdly, by the reduction of properties measured in complex units such as momentum and pressure to more basic dimensions such as mass, length and time on the basis of the two previous approaches. In this last way, all complex units can be re-expressed in terms of a few basic properties between which no fixed relationship is recognised. This recognition is not necessary provided that a sufficiently small group of dimensions is involved. For example, a general relationship between mass and time or for that matter between mass and length is not recognised. In fact, what have been called fundamental dimensions could be defined as those dimensions between which it has not been possible to establish a fixed universal relationship. It is necessary however to realise that no absolute importance can be attached to these particular dimensions, since another set could equally well be chosen and all complex unit phenomena expressed in terms of this second set. For example, length need not have been chosen as a fundamental property and volume chosen instead. In which case, length would now appear to be a more complex phenomena being volume raised to the power of a third, - which thus alters the concept of length considerably. Anthropomorphic considerations alone have established length as a primary dimension.

8.4.4.5 Constants

Now the remaining aspect of the construction of an equation is the numerical or constant. Wherever numbers are used in the mathematical interpretation of a law they must always mean numbers of things. Numbers without an associated unit mean absolutely nothing, so all the constants employed in modelling laws have a physical reality or at least a quasi-physical description. This is true not only for the constants used to correct zero errors (e.g. 273 in temperature measurements) where it is obvious what the constant is a measure of, i.e. the unused part of a measuring scale but also for powers where dimensions have been multiplied by themselves and to type constants which are called at the moment, scale factors. These are not just numbers, they are numbers of things and must always be regarded as such.

8.4.5 Law Structure

8.4.5.1 Function

It is appropriate at this point to reconsider the role of a law. At the beginning of this note it was put that the function of a mathematically expressed law was to represent cause and effect relationship. At first sight, laws do not appear to serve this purpose but merely be statements of state such as, if this property of an object measured in *a* units is 30 then another property measured in *b* units will be 60. This infers that the length of a block of wood could be measured which was, say, sixteen inches and then its temperature measured and found to be say 64°F and it seems reasonable to make the statement that - for this block of wood in these circumstances 16 inches is equivalent to 64°F or that $L = 0.25T$ where L is length and T is temperature.

Now obviously, these are not very useful statements to make and are not what is required and certainly are not what is normally meant by a 'law'. What is of real concern are the relationships between whole series or families of states.

It is desirable to know how observed objects or entities respond to changes in state and the laws that are of interest are really laws of change as opposed to laws of state. So it is useful to restate the law equation (8.52) as:-

$$F(Xa) - F(Xb) = F(Ya) - F(Yb)$$

This does not take the argument much further forward since it is not known if the functions of X and Y for one change are the same as those which are required to represent another. This is not an insurmountable difficulty since if, the functions of X and Y change continuously they can, in principle, be transformed into more complex functions which take account of this continuous change. If, however, the functions change discontinuously, it is unreasonable to expect to produce a law which infers continuous behaviour. Therefore a single discontinuity requires two separate laws as the analogues for the two physical characteristics being represented. Such a discontinuity is synonymous with the hierarchical boundary mentioned in the Introduction.

8.4.5.2 Law Class

In order to analyse the relationships between the mathematical statement of a law and its physical manifestation, it is useful to use a simple system of classification of the mathematical statements based on the degree of similarity between the scales of measurement used for the cause and that used for the effect in a given situation. For example, a law in the first and simplest class might be relating the length of one side of a right-angled triangle to another, when both lengths are measured using the same ruler. In the second class, one side would be measured with an imperial ruler and the other with a metric ruler. A law in the third class might relate the length of the base of a fixed height right-angled triangle to its area, where there was a special device to measure area directly and so on up to the last class of multi-parameter laws where each parameter is quantifiable only on complex unit scales.

8.4.5.3 Class 1 Laws

These laws represent situations where causal behaviour can be measured in single dimensions with identical scales. Consider the triangle shown in Fig. 8-5 for which the law can be used :-

$BC = AB \sin BAC$

if for example BAC is 30° then $\sin BAC = 0.50$ and the law can be written as:-

$BC = 0.50\ AB$

and in a particular case where $BC = 20$ ft and $AB = 40$ ft

$$20 \text{ ft} = 0.50 \times 40 \text{ ft} .$$

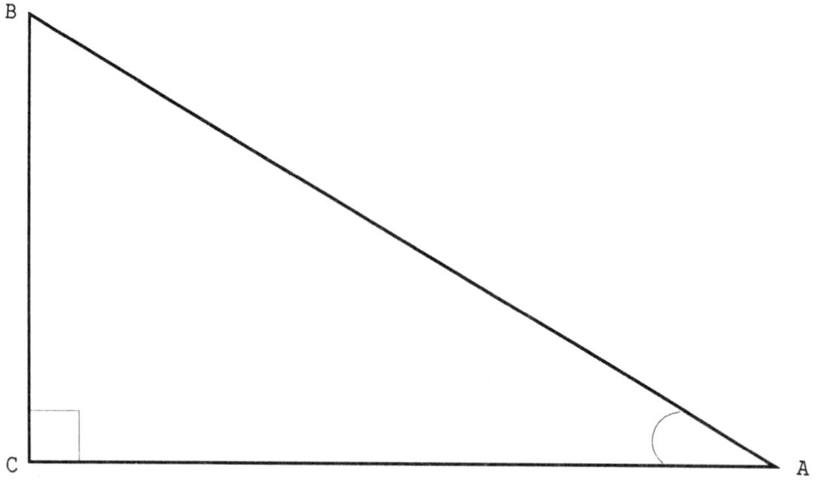

Fig. 8-5. For the demonstration of single dimension laws

It appears that when the law is written in this form, it contains a number, ie 0.50, which has no associated unit; a situation which has already been declared inadmissible cf section on constants. The only solution to the problem is to rewrite the equation through the following transformations:

$$1 \times (20 \text{ ft}) = 0.50\ (40 \text{ ft})$$

since half a unit is equally inadmissible then:-

$$2 \times (20 \text{ ft}) = 1 \times (40 \text{ ft})$$

This has reduced the law to a simple statement of correspondence between measurements on two new length scales (one with 20 ft units and the other with 40 ft units): that a length of 1 unit on one scale is identically equal to a length of 2 units on the other scale. It is only when the law is stated in this form that the valid use of an equals sign can be claimed, since it is in this form that the correspondence between the law and the physical operation required for its justification becomes obvious.

8.4.5.4 Class 2 Laws

These laws also represent situations where causal behaviour can be measured in a single dimension however in this case with linearly related scales of differing unit sizes. If the previous example is reformed using Metric and Imperial scales, then a similar set of transformations will produce a mathematical form for the law which is identical to that found for the class 1 laws.

8.4.5.5 Class 3 Laws

These laws concern relationships between dissimilar properties such as length and area or voltage and current. To demonstrate how homogeneity can be obtained in this type of law consider this last example, i.e. Ohm's Law. This law was originally stated in the following form:-

> The current through a conductor, under constant conditions is proportional to the difference of potential across the conductor.

It is more familiarly written nowadays as:-

$V = IR$

where

V is the potential difference across the conductor measured in volts
I is the current through the conductor measured in amperes
R is the so-called resistance of the conductor measured in ohms.

The data from Ohm's original work was in terms of V and I only and the law to him would in modern conventional form look like this:-

$I = CV$,

where C is an empirical constant.

Since within the limits of experimental accuracy this law is demonstrably correct, it must be possible to demonstrate the equivalence of measurements made on different scales of different units. To accomplish such a demonstration the law has to be taken through a series of transformations.

Let n represent a number, a the units of current and v the units of potential difference. Substituting these to represent a particular solution to the equation of the law:-

$$n_1 a = c n_2 v$$

As explained in the section on constants, they are never just numbers but always numbers of things (units). So the law can again be rewritten as:-

$$n_1 a = n_3 (c \text{ units}) n_2 v$$

Now the only ways that this form of the law can be validated is for the dimension equivalence:-

$$a \equiv c \text{ units } v$$

or alternatively

$$c \text{ units} \equiv \left(\frac{a}{v}\right) \text{ to be true.}$$

Therefore the law can be put down as:-

$$n_1 a = n_4 a / n_5 v \cdot n_2 v$$

and

$$n_1 a / n_4 a = n_2 v / n_5 v.$$

In conventional terms such an equation is termed non-dimensional; a description which is unsatisfactory - even if it has had its uses. What has happened however, is that both sides of the law have been written in terms of a unit *common* to each, in exactly the same way as the length unit of the Class 1 law, example given earlier. There now only exists a scale difference in the law which means it is now in Class 1 and can be dealt with as detailed above. Now the fascinating question arises - in what dimension is the law cast? The answer can be found by reviewing the choices to be made prior to making a measurement of change. These are:-

1. A unit type, i.e. dimension - length, time etc.
2. A scale.
3. A unit size.

As far as the object or situation is itself concerned these choices are arbitrary. The dimension in which the measurements are made may not be the most appropriate to describe the change. The scale may not conform to the scale relating the internal units of change of the system such as molecule size, crystal size, dislocation spacing etc. Finally, the unit size may not conform with the internal unit of change of the system, e.g. if one tries to relate the number of stairs required to reach from one storey of a building to the next, either the storey height must be measured in the same units as each stair or vice versa.

In setting up a law to represent causal changes in a system, at the very least, two such groups of choice must be made; a further piece of information is however available, i.e. a correlation between the two sets of measurements does exist (either by assumption or observation).

As described earlier, a measurement requires an operational technique for it to be made. This technique in itself prescribes the level of discrimination, that is, the smallest detectable unit of change: So the initially arbitrary choice of unit size is modified by taking some fraction of it as the real unit [Section 8.4.4.1].

The two scales can be made to conform with each other firstly, by added constants to correct (if necessary) for zero errors, secondly by the use of multiplying constants make linearly related scales conform and lastly by the use of powers which reduce other types of scale to linearly related ones which are then treated as before. Finally, the two dimensions chosen to describe the change must demonstrably conform. As stated earlier, interactions in systems can only take place at a common level which means in turn, that system changes which are channelled through that level must be expressible in a common dimension. This is exactly the function performed by money in society when, for example, it is used by a carpenter to convert some of his own work (as measured by his earnings) into bricklayer's work, by paying money the latter that is earned by the former. In this case, a dimension of equivalence has been invented. Every law must be either implicitly or explicitly (i.e. when considered non-dimensional!) be expressed in this dimension.

8.4.5.6 Higher Class Laws

Laws in classes higher than those analysed so far can be treated in a similar manner. It is not a particular useful exercise to go through them all here.

8.4.6 Discontinuities

The causes of discontinuities in the mathematical description of a natural phenomenon are twofold. Firstly, there are discontinuities in behaviour which are a direct result of the interaction between an object and its environment. These discontinuities are uncontrollable mathematically and can only be affected by a change in characteristic of either the object or its environment. For example, consider a compression test on a concrete cube, it is possible to produce a simple law to represent the load-deformation history of such a test that is until the cube crushes at its maximum load and there is an obvious change in the mechanism by which the concrete resists its load. It would be possible to control this change in behaviour by altering the cube environment to a triaxial stress state. Secondly, there can be discontinuities in the mathematical descriptions of behaviour which are directly attributable to implicit assumptions concerning the rules of mathematics, made in the construction of a law.

The important rules that break down are:-

1. Commutative law of addition

$a+b = b+a$

2. Associative law of addition

$(a+b)+c = a+(b+c)$

3. Law of subtraction

$a+x = b$ has a unique solution $x = b-a$

4. Commutative law of multiplication

$ab = ba$

5. Associative law of multiplication

$(ab)c = a(bc)$

6. Law of division

$ax = b$ $(a=0)$ has a unique solution $x = b$

7. Distributive law

$(a+b)c = ac+bc$

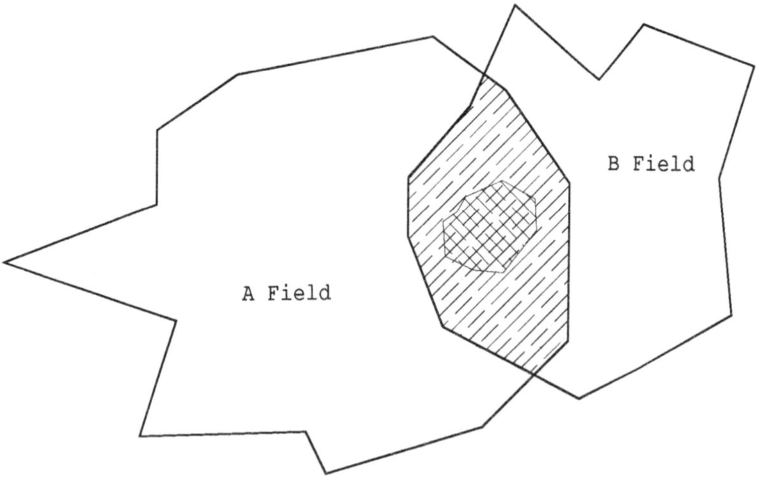

Fig. 8-6. Interacting measurement fields

These rules describe the nature of the number fields available to model an empirical observation. Any lack of conformity between these rules and those which control the relationship between empirical measurements will lead to an **apparent** discontinuity in the behaviour of the observed system. Likewise, since the number of object units involved in a change, ie constitutive elements must be matched by the number of measured elements field assumed in the mathematics in order that a discontinuity in behaviour should not apparently exist. It is interesting to note that measuring too accurately is just as likely to expose unwanted discontinuities as measuring too coarsely.

8.4.7 The Representation of Empirical Data by a Simple Law Form

Consider the two measurement potential fields A and B shown in Fig. 8-6. This figure represents the complete range of measurements obtainable through the chosen operational technique.The shaded area, indicating the zone of overlays, represents the region where there is a physically identifiable causal link between measurements A and B. As an illustration of these fields, take the case of temperature and length measurements of a sample of steel. The boundary of the A field representing say, temperature measurements and that of the B field representing length measurements both indicate the units beyond which the characteristics of temperature and length become unmeasurable either through limitations of the measurement technique or because they cease to exist. If, for example, the temperature of the steel is raised to 3000°C, it will melt and its length will not be a measurable property or alternatively at the lower temperature boundary of absolute zero, a variety of changes could occur without any corresponding measurable temperature changes. Returning to the general case of two sets of empirical data being taken from a changing system, Figure 8-7 shows the four possible classes of data set string available without crossing the boundary of the zone of overlap. Data sets which do cross the zero boundary necessarily involve discontinuities in behaviour such as the stress-strain curve for mild steel (Fig. 8-8). The four classes are, in reverse order: Class 4, where both ends of the data string coincide with the *A* and *B* field boundaries, Class 3, where both ends of the data string coincide with one field boundary, Class 2 where one end of the string coincides with one field boundary and Class 1, where there is no simple relationship between the data string position and the field boundaries. If a given data string can be classified in any of these four ways then each pair of measurements will be unique and no strong discontinuities will exist in the string. This is not sufficient to produce a **simple** continuous law however and further restrictions are required to avoid weak discontinuities. These relate to the corresponding fields of 1st, 2nd and so on differences, as shown in Fig. 8-9. The number of these fields which are relevant is easily deduced since the units of each are identical to those of the prime fields and consequently the level of discrimination relates to the two operational techniques of measurement that will control the smallest detectable difference. When the differences in one of the fields is below the level of discrimination, they become irrelevant to the law.

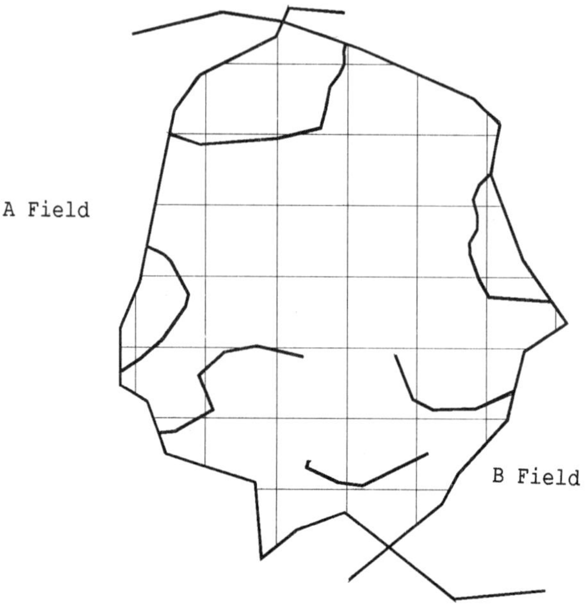

Fig. 8-7. Possible classes of string for arbitrarily collected data

This leads to the following situation:-

Measurements $A_m \equiv B_m$,

where m is the number of pieces.

1st Differences of Measurements of data available. $\delta A_m \equiv \delta B_m$

2nd Differences of Measurements $\delta^2 A_m \equiv \delta_2 B_m$

3rd Differences of Measurements $\delta^3 A_m \equiv \delta_3 B_m$

*n*th Differences of Measurements $\delta^n A_m \equiv \delta_n B_m$

For continuity each of these equivalences must be unique.

The classes of data string which the differences can form in their respective fields are identical to those described earlier, but are not necessarily the same for each stage of the processing. Certain information about the difference data strings is automatically available. Firstly, the units (dimension) in which the differences are measured are the same as those of the prime measurements, A and B. Secondly, there is a change in the operational technique used for obtaining the data strings between the prime measurements and the first differences although the new technique is maintained for the future development of data in the form of higher differences. Thirdly, the scales are the same as for the prime field measurements. If there are insufficient empirical data available, it may be impossible to arrive at sufficiently high order differences which are less than the levels of discrimination - in these cases it will be impossible to judge whether a continuous law is appropriate.

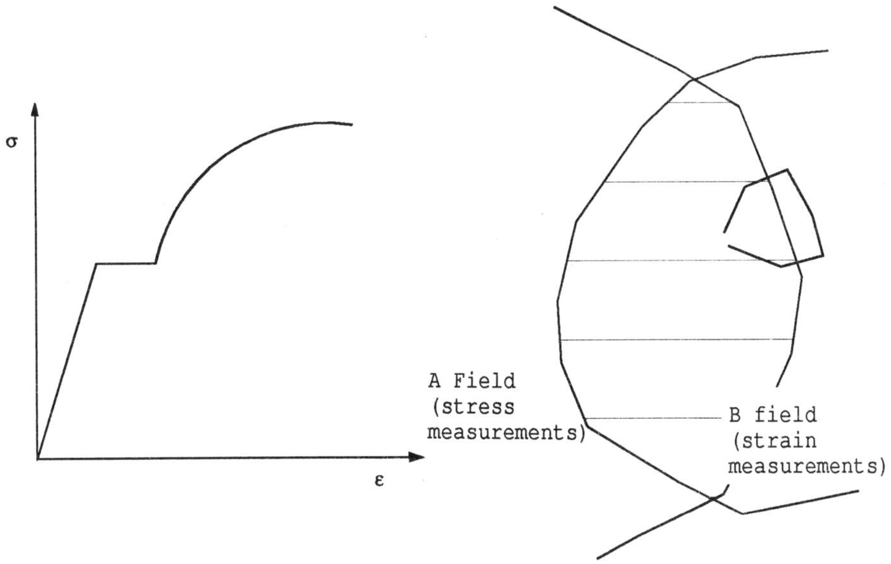

Fig. 8-8. Stress-strain curve for mild steel with corresponding data set mapped onto *A* and *B* measurement potential fields

This discussion has so far, been concerned with an investigation of the empirical data available from an experiment. The statement of a law to represent such data generally implies the existence of a considerably greater body of data namely that measurements have been made at single unit increments over the full range of data acquisition and that continuity of system behaviour exists between actual data points. This can be stated formally thus:-

Since a simple law is being sought, the earlier discussion indicates that zero corrections of the data must be applied.

This preamble has identified some ground rules which must be satisfied if empirical data is to be represented by a simple mathematical statement or law. These rules are concerned partly with the characteristics of the observed subject and partly with the characteristics of the observer. It is implicit that the laws so generated are not natural or absolute.

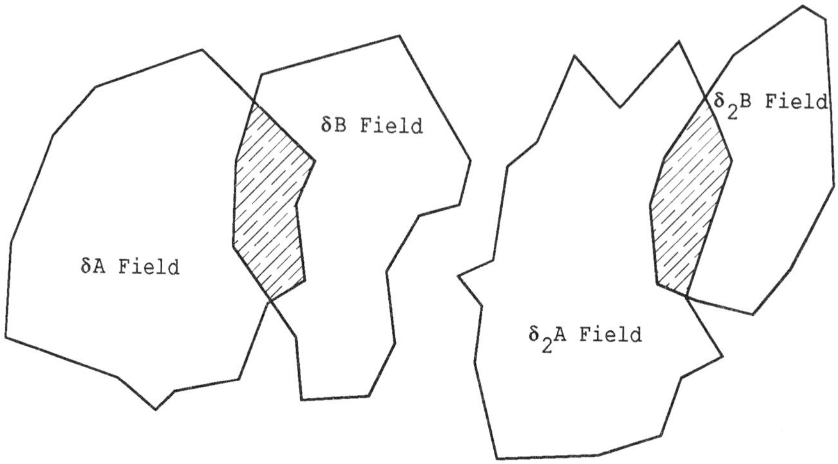

Fig. 8-9. First and second difference fields

In summary, the response of the subject to its changing environment must reflect a continuity of response mechanisms and at least one boundary state, where there is a significant change in mechanism, must be identified. The observer's measurements must have a minimum level of discrimination which will **not** provide information on small scale responses which, by their complexity, will overshadow the simple large scale behaviour to be modelled. Also the data must be related to the value recorded at the major discontinuity in behaviour of the subject identified during testing.
With these points in mind, the numerical law:-

$$f(x,y,z,...) = \text{const.}$$

which is expressed with respect to classes of similar scales [*X*], [*Y*], [*Z*],... It can be demonstrated that the function *f* must be of the form

$$x^m y^n z^o ... = \text{const.}$$

Differing proofs of this statement have been given by Ellis [1966] and Bridgman [1931].

9

REFERENCES

Chapter 1

Chilver, A. H.	1967	Thin walled structures. Chatto and Windus. London
Descartes, R.	1968	Discourse on method and meditation. Translated by F. E. Sutcliffe, Penguin Books, London
Granger, F.	1934	Vitruvius on architecture. Heinemann, London
Griffiths, H., Pugsley, A. and Saunders, O.	1968	Report of the inquiry into the collapse of flats at Ronan Point, Canning Town. HMSO, London
ISE	1988	Stability of buildings. The Institution of Structural Engineers, London
Komarakul-na-nakorn, A. and Arora, J. S.	1990	Stability criteria: A review. Computers and Structures. Vol. *37*, No1, 35-49

Chapter 2

Armer, G. S. T. and Grimer, F. J.	1989	On the use of the full stress-strain characteristics in structural analysis. Materials and Structures. Vol. *22*, 48-52
Bate, S. C. C.	1974	Report on the failure of roof beam at Sir John Cass's Foundation and Red Coat Church of England secondary school Stepney. BRE Current Paper CP58/74. Building Research Establishment, Watford, UK
BRE	1973	Report on the collapse of the roof of the assembly hall of the Camden School for girls. HMSO, London
Griffiths, H., Pugsley, A. and Saunders O.	1968	Report of the inquiry into the collapse of flats at Ronan Point, Canning Town. HMSO. London
ISE	1988	Stability of buildings. The Institution of Structural Engineers, London
Mainstone, R. J.	1959	Structural theory and design. 1. Early propor-

tional theories. Architecture and Building, Vol. *34* (3), 106-113

Morgan, B. G. 1961 Canonical design in English mediaeval architecture. Liverpool University Press

Chapter 3

Ackroyd, T. N . W. 1981 Progressive collapse revisited. (The problem of structural failure.) The Structural Engineer, Vol. *59a*, No. 12, 387-391

Andronov, A. A. and Pontriagin, L. S. 1937 Grubye sistemy. Doklady Akademii nauk SSSR, Vol. *14*, No. 5, 247-51

Bennett, R. M. 1988 Formulations for probability of progressive collapse. Structural Safety. Vol. *5*, 67-77

BRE 1985 The structure of Ronan Point and other Taylor Woodrow-Anglian buildings. BRE Report, BR 63. Building Research Establishment, Watford, UK

BRE 1986 Large panel system dwellings: Preliminary information on ownership and condition. BRE Report, BR 74. Building Research Establishment, Watford

BRE 1987 The structural adequacy and durability of large panel system buildings. BRE Report, BR 107. Building Research Establishment, Watford, UK

BSI 1985 The structural use of masonry, BS 5628: Part 1. British Standards Institution, London

BSI 1985 The structural use of concrete, BS 8110: Part 1. British Standards Institution, London

BSI 1990 The structural use of steel. British Standards Institution, London

Byrd, T. 1981 Ronan Point: The final verdict. New Civil Engineer. July, 15-16

Currie, R. J., Armer, G. S. T., Reeves, B. R. and Moore, J. F. A. 1987 The structural adequacy and durability of large panel buildings. Building Research Report, DOE, Watford, UK

DOE 1974 Collapse of roof beams. Circular Letter BRA/1086 2, (May). Department of the Environment, London

Frangopol, D. M. and Klisinski, M. 1989 Material behaviour and optimal design of structural systems. Journal of Structural Engineering Vol. *115*, No. 5

Griffiths, H., Pugsley, A. and Saunders, O. 1968 Report of the inquiry into the collapse of flats at Ronan Point, Canning Town. HMSO, London

Hirst, M. J. S. 1974 The structural planform of large panel buildings and design against progressive collapse. Build International. Vol. *7*, 253-263

Hunt, G. W. 1982 Elastic stability: In structural mechanics and applied mathematics. Collapse: The buckling of structures in theory and practice. Eds: J. M. T. Thompson and G. W. Hunt. Cambridge University Press

Leyendecker, E. V. and Burnett, E. F. P. 1976 The incidence of abnormal loading in residential buildings. NBS Building Science Series *89*, US

		Department of Commerce
Leyendecker, E. V. and Ellingwood, B. R.	1977	Design methods for reducing the risk of progressive collapse in buildings. NBS Building Science Series *98*. US Department of Commerce
Leyender, E. V. and Fattal, S. G.	1977	Investigation of the Skyline Plaza collapse in Fairfax County, Virginia. NBS National Building Science Series *94*. Washington, USA
Mainstone, R. J.	1971	The breakage of glass windows by gas explosions. BRS Current Paper, CP26/71. Building Research Establishment, Watford, UK
Mainstone, R. J.	1973	The hazard of internal blast. BRE Current Paper, CP11/73. Building Research Establishment, Watford, UK
Massonet, C. E.	1982	The collapse of struts, trusses and frames: A survey of up-to-date problems. Collapse: The buckling of structures in theory and practice. Eds: J. M. T. Thompson and G. W. Hunt. Cambridge, University Press
Morton, J.	1985	Accidental damage, robustness and stability. Brick Development Association, UK
NCE	1985	Ulster tests confirm ceiling overload. New Civil Engineer, 6th June.
Pekau, O. A., Zielinski, Z. A. and Hum, D.	1988	Dynamic effects of panel failure in precast concrete shearwalls. ACI Structural Journal, May/June.

Chapter 4

Armer, G. S. T. Fewell and Mayne, J. R.	1987	The use of partial factors in structural design. Symposium Proc. Building Research Establishment. Watford, UK
Armer, G. S. T.	1988	Structural safety - some problems of achievement and control. Second century of the skyscraper. Van Nostrand Reinhold Co. New York, USA
Bridgman, P. W.	1931	Dimensional analysis. Yale University Press, New Haven, USA
Ellis, B.	1966	Basic concepts of measurements. Cambridge University Press, UK
Ellis, B. R. and Littler, J. D.	1988	Dynamic response of nine similar tower blocks. Journal of Wind Engineering and Industrial Dynamics. Vol. *28*, 339-349
Moore, J. F. A.	1991	Construction monitoring. Blackie and Sons. Glasgow, UK
Thom, R.	1975	Structural stability and morphogenesis: An outline of a general theory of models. W. A. Benjamin Inc. Reading, Massachusetts. USA
Wood, R. H.	1955	The absolute and the intuitive in dimensional analysis. Euler Society. University College, London.

Chapter 5

Armer, G. S. T.	1983	The stability of structures. Building Research and Practice. The Journal of CIB. *12*, (4), July/August, 216-221
Armer, G. S. T. and Grimer, F. J.	1972	The stability of large complex structures. Building Research Station, IN 112/72
Asztalos, Z.	1972	Buckling of single-bay, multi-storey frameworks by the continuum method, taking column shortening into consideration (in Hungarian). Magyar Épitöipar. 471-474
Beck, H.	1956	Ein neues Berechnungsverfahren für gegliederte Scheiben, dargestellt am Beispiel des Vierendelträgers. Der Bauingenieur. *31*, 436-443
Beck, H.	1959	Ein Beitrag zur Berücksichtigung der Dehnungsverformungen bei Rahmen mit schlanken und gedrungenen Konstruktionsgliedern. Die Bautechnik. *36*, 178-184
Bleich, F.	1952	Buckling Strength of Metal Structures McGraw-Hill, New York.
Chwalla, E.	1938	Die Stabilitat Lotrecht Belasteter Rechteckrahmen. Der Bauingenieur. *19*, Heft 5/6, 69-75
Council on Tall Buildings	1978	Planning and design of tall buildings, a Monograph in 5 Volumes. ASCE, New York
Croll, A. G. J. and Walker, C. A.	1972	Elements of structural stability. London, Macmillan
Csonka, P.	1950	Procedure for rectangular sway frames (in Hungarian). Mérnöki Továbbképzö Intézet, Épitészet *39*, Tudományos Könyvkiadó
Csonka, P.	1956	Uber proportionierte Rahmen. Die Bautechnik. *33* Jahrgang, Heft1, 19-20
Csonka, P.	1961	Buckling of bars elastically built-in along their entire length. Acta Technica. Tomus *XXXII* Fasciculi 3-4, 423-427
Dulácska, E. and Kollár, L.	1960	Angenäherte Berechnung des Momentenzuwachses und der Stabilität von gedrückten Rahmenstielen. Die Bautechnik. *37*, (3), 98
Glück, J., Goldberg, J., Wang, C. et al.	1979	Stability - in "Planning and design of tall buildings." SB-4, 238-342, ASCE, New York
Goldberg, J. E.	1973	Approximate methods for stability and frequency analysis of tall buildings. Regional Conference on "Planning and Design of Tall Buildings," 123-146, Madrid
Goschy, B.	1990	Design of buildings to withstand abnormal loading. Butterworths, London
Hegedüs, I. and Kollár, L. P.	1984a	Buckling of sandwich columns with thin faces under distributed normal loads. Acta Technica Hung. *97,* 111-122
Hegedüs, I. and Kollár, L. P.	1984b	Buckling of sandwich columns with thick faces subjected to axial loads of arbitrary distribution. Acta Technica Hung. *97,* 123-132
Hegedüs, I. and	1987	Stabilitätsuntersuchung von Rahmen und Wand-

Kollár, L. P.		scheiben mit der Sandwichtheorie. Die Bautechnik. *64*, 420-425
Horne, M. R. and Merchant, W.	1965	The stability of frames. Pergamon, Oxford
Kollár, L.	1967	Bestimmung der horizontalen Eigenfrequenz von Hochhaus-Stockwerkrahmen. Die Bautechnik. *44*, 385-388
Kollár, L.	1972	Stability of frame and column systems (in Hungarian. Budapesti Városépitési Tervezö Vállalat. Budapest
Kollár, L. and Hegedüs, I.	1985	Analysis and design of space frames by the continuum method. Akadémia Kiadó, Budapest
Kollár, L. Ed.	1991	Special stability problems in civil engineering (in Hungarian). Akadémia Kiadó
Kollár, L. P.	1986	Stability analysis of frameworks and coupled shear walls by the continuum method and by the method of finite differences (in Hungarian) CSc (PhD) Thesis. Hungarian Academy of Sciences
Rosman, R.	1972	The continuum analysis of shear-wall structures. Proceedings of the International Conference on the Planning and Design of Tall Buildings. Bethlehem, Pennsylvania
Sattler, K.	1953	"Das Durchbiegeverfahren" zur Lösung von Stabilitätsproblemen. Die Bautechnik. *30*, 288-294 and 326-331
Sigalov, E.	1957	Free vibration of frame- coupled shear wall systems (in Russian). Trudi MIIGSZ. Szbornik szedmoi. Gosztroizdat
Szerémi, L.	1978	Stiffening system of multistorey buildings by the continuum model. Periodica Politechnica, Civil Engineering. *22*, 205-218
Szittner, A. et al	1979	Small-scale tests of frameworks (in Hungarian). Technical University of Budapest, Department of Steel Structures. No. 202002/79
Timoshenko, S. and Gere, J.	1961	Theory of elastic stability. New York, McGraw-Hill.
Wang, C. K.	1973	Computer methods in advanced structural analysis. Intext Educational Publishers, New York.
Weaver, W. Jr and Gere, J. M.	1990	Matrix analysis of framed structures. Van Nostrand Reinhold
Wood, R. H.	1974	Effective lengths of columns in multi-storey buildings. The Structural Engineer, *52*. In three parts: 235-244, 295-302 and 341-346
Wood, R. H.	1975	Effective lengths of columns in multi-storey buildings (discussion). The Structural Engineer, 235-241
Zalka, K. A.	1976	Stability investigation of frameworks loaded at the top by concentrated forces, with the continuum method. Acta Technica Hung. *83*, 357-374
Zalka, K. A.	1977a	Stability investigation of frameworks subjected to distributed normal loads, with the aid of the continuum method. Acta Technica Hung. *84*, 43-59

Zalka, K. A. 1977b Stability analysis of frameworks by directly taking into account the sway access of the columns. Acta Technica Hung. *84*, 307-314

Zalka, K. A. 1987 Stability of large frameworks on hinged supports. Acta Technica Hung. *100* (3-4), 355-383

Zalka, K. A. 1991 On the accuracy of the continuum method of frameworks (in Hungarian). Építés- Építészettudomány, Budapest. (Accepted for publication)

Zienkiewicz, O. C. 1967 The finite element method in structural and continuum mechanics. McGraw-Hill, London New York

Chapter 6

Allen, H. G. 1969 Analysis and design of structural sandwich panels. Oxford

Anderson, D. and Lok, T. S. 1983 An elasto-plastic hand method for unbraced rigid-jointed steel frames. Granada, London. Proceedings of the Michael R. Horne Conference: Instability and plastic collapse of steel structures. Edited by L. J. Morris

Beck, H. 1972 Stability in the design of tall concrete buildings. International Conference on Planning and Design of Tall Buildings, Lehigh University, Pennsylvania, Vol. *III*, 499-516

Beck, H. and König, G. 1966 Restraining forces in the analysis of tall buildings. Symposium on tall buildings, Proceedings. Oxford, Pergamon Press, 37-53

Croll, A. G. J. and Walker, C. A. 1972 Elements of structural stability. London, Macmillan

Fazakas, M., Kaján, L, Soós, Zs. and Zalka, K. A. 1988 Planar buckling of frameworks, shear walls, coupled shear walls and cross wall system buildings. Computer procedure. Budapest

Gáspár, Zs. 1976 Elastic analysis of structures developing large deformations (in Hungarian). CSc (PhD) Thesis. Hungarian Academy of Sciences

Hegedüs, I. 1987 Comment on "Stability of large frameworks on hinged supports," a study by K. A. Zalka. Acta Technica Hung. *100*, 385-386

Hegedüs, I. and Kollár, L. P. 1984a Buckling of sandwich columns with thin faces under distributed normal loads. Acta Technica Hung. *97,* 111-122

Hegedüs, I. and Kollár, L. P. 1984b Buckling of sandwich columns with thick faces subjected to axial loads of arbitrary distribution. Acta Technica Hung. *97,* 123-132

Hegedüs, I. and Kollár, L. P. 1988a Stability analysis of bars elastically restrained from rotation along their entire length. Acta Technica Hung., Civil Eng. *101*, 57-65

Hegedüs, I. and Kollár, L. P. 1988b Generalized bar models and their physical interpretation. Acta Technica Hung., Civil Eng. *101*, 67-93

Horne, M. R. 1963 Elasto-plastic failute loads of plane frames.

Proceedings of the Royal Society, *A274*, 343-364

Horne, M. R. 1981 Discussion on "Generalization of the Dunkerley theorem and its application in the theory of elastic stability" by T. Tarnai. Manuscript

Howlett, J. H., Jenkins, W. M. and Stainsby, R. Ed. 1981 Joints in Structural Steelwork. Pentech Press, London

Irwin, A. W. 1984 Design of shear wall buildings. Construction Industry Research and Information Association

Jones, S. W., Kirby, P. A. and Nethercot, D. A. 1979 The analysis of frames with semi-rigid connections - a-state-of-the-art report. Department of Civil and Structural Engineering, University of Sheffield

Kollár, L. P. 1986 Stability analysis of frameworks and coupled shear walls by the continuum method and by the method of finite differences (in Hungarian). CSc (PhD) Thesis. Hungarian Academy of Sciences

Kollár, L. P. 1988 Buckling analysis of coupled shear walls by the multi-layer sandwich model. Acta Technica Hung. *99*, 317-332

König, G., Ambos, G, Bubenheim, H. J. and Heunisch, M. 1972 Stability of frames and coupled shear walls. International Conference on Planning and Design of Tall Buildings, Vol. *III*, 613-620. Lehigh University, Bethlehem, Pennsylvania

MacLeod, I. A. 1967 Lateral stiffness of shear walls with openings. in Tall Buildings. Edited by A. Coull and B. Stafford Smith. Pergamon Press, 223-244

Rosman, R. 1965a Die statische Berechnung von Hochhauswänden mit öffnungsreihen Bauengenieur-Praxis, Heft *65*. Wilhelm Ernst & Sohn, Berlin

Rosman, R. 1965b Windscheiben mit beliebig vielen öffnungsreihen Deutsche Bauzeitschrift. Heft *7*

Rosman, R. 1972 The continuum analysis of shear-wall structures International Conference on the Planning and Design of Tall Buildings. Lehigh University, Bethlehem, Pennsylvania

Rosman, R. 1973 Dynamics and stability of shear wall building structures. Proceedings of the Institution of Civil Engineers. *55*, 411-423

Schueller, W. 1977 High-rise building structures. John Wiley and Sons Inc.

Stafford Smith, B. and Coull, A. 1972 Elastic analysis of tall concrete buildings. International Conference on Planning and Design of Tall Buildings, Vol. *III*, 157-170. Lehigh University, Bethlehem, Pennsylvania

Szittner, A. et al 1979 Small-scale tests of frameworks (in Hungarian). Technical University of Budapest, Department of Steel Structures. No. 202002/79

Taranath, B. S. 1988 Structural analysis and design of tall buildings. McGraw-Hill, London

Tarnai, T. 1981 Generalization of the Dunkerley theorem and its application in the theory of elastic stability. 9th International Congress on the Application of Mathematics in Engineering Science, Weimar.

Wood, R. H. 1974 Effective lengths of columns in multi-storey buildings. The Structural Engineers, *52*. In three parts: 235-244, 295-302 and 341-346

Zalka, K. A. 1979 Buckling of a cantilever subjected to distributed normal loads, taking the shear deformation into account. Acta Technica Hung. *89*, 479-508

Zalka, K. A. 1982a Stability analysis of coupled shear walls by the continuum approach (in Hungarian). Épités-Épitészettudomány, 143-154

Zalka, K. A. 1982b Stability analysis of the support system of buildings. CSc (PhD) Thesis. Budapest

Chapter 7

Akesson, B. 1980 Overall buckling in bending and torsion of rack structures. Thin-walled structures, edited by J. Rhodes and A. C. Walker, 127-144

Beck, H. and Schäfer, H. 1969 Berechnung von Hochhäusern durch Zusammenfassung aller aussteifenden Bauteile zu eine balken. Der Bauingenieur. *44*, 80-87

Bornscheuer, F. W. 1952 Systematische Darstellung des Biege- und Verdrehvorganges unter besonderer Berücksicht. Der Stahlbau, *21*. Jahrgang, Heft 1, 1-9

Brandt, B., Schäfer, H. and Reeh, H. 1975 Zum Stabilitätsnachweis von Hochhäusern. Beton und Stahlbetonbau, 9/1975, 211-223

Danay, A., Gellert, M. and Glück, J. 1975 Continuum method for overall stability of tall asymmetric buildings. Proc. of ASCE, Journal of the Structural Division, *101*, ST12, 2505-2521

Fazakas, M., Kaján, L., Soós, Zs. and Zalka, K. A. 1989 Spatial stability analysis of multistorey buildings. Computer procedure. Budapest

Glück, A. and and Gellert, M. 1972 Buckling of lateral restrained thin-walled cantilevers of open cross section. Proc. of the American Society of Civil Engineers, Journal of the Structural Division, *98*, 2031-2042

Harman, D. and Walker, W. H. 1975 Analysis of tall buildings using member groups. Proceedings of ASCE, Journal of the Structural Division, *101*, 567-583

Horacek, E. Räumliche Scheibensysteme mit Öffnungsreihen bei Hochbauten. Der Bauingenieur
1970 Teil I: *45*, 447-452
1971 Teil II: *46*, 400-404
1974 Teil III: *49*, 77-83

Irwin, A. W. 1984 Design of shear wall buildings. Construction Industry, Research and Information Association

Kókai, T. 1983 Axis of twist for the support system of tall buildings (in Hungarian). University doctoral thesis. Department of Reinforced Concrete Structures, Budapest Technical University

Kollár, L. 1977 Stiffening of buildings against torsional buckling (in Hungarian). Magyar Épitöipar, 150-154

Kollár, L. P. 1986 Stability analysis of frameworks and coupled shear walls by the continuum method and by the method of finite differences (in Hungarian). CSc (PhD) Thesis. Hungarian Academy of Sciences

Nair, R. S. 1975 Overall elastic stability of multistorey buildings. Proceedings of ASCE, Journal of the Structural Division, *101*, (ST 12), 2487-2503

Pearson, C. E. 1956 Remarks on the centre of shear. Z. angew. Math. Mech. Bd. 36, Nr *314*, 94-96

Podolsky, D. M. 1970 Spatial analysis of tall buildings (in Russian) Structural Mechanics. *68*, 63-69. Moscow

Rosman, R. 1965 Die statische Berechnung von Hochhauswänden mit öffnungsreihen. Bauengenieur-Praxis, Heft *65*, Wilhelm Ernst & Sohn, Berlin

Rosman, R. 1973 Dynamics and stability of shear wall building structures. Proc. of the Institution of Civil Engineers. *55*, 411-423

Rosman, R. 1980 Stabilität im Grundriβ unsymmetrischer Stützen- und Wandscheibensysteme. Die Bautechnik, *1/1980* 21-32

Stüssi, F. 1965 Die Grenzlagen der Schubmittelpunktes bei Kastenträgern. IVBH Publications *25*, 279-315

Szmodits, K. 1975 Handbook for the design of panel type structures (in Hungarian). Építéstudományi Intézet, Budapest

Taranath, B. S. 1968 The torsional behaviour of open section shear wall structures - Thesis for PhD, University of Southampton

Taranath, B. S. 1988 Structural analysis and design of tall buildings. McGraw-Hill, London

Timoshenko, S. and Gere, J. 1961 Theory of elastic stability. New York, McGraw-Hill

Vlasov, V. Z. 1961 Thin-walled elastic beams. Israeli Program for Scientific Translation, Jerusalem

Yarimci, E. 1972 The equivalent-beam approach in tall building stability analysis. Proceedings of the International Conference on the Planning and Design of Tall Buildings. Vol. *II-16*, 553-563

Zalka, K. A. 1980a Torsional buckling of a cantilever subjected to distributed normal loads. Acta Technica Hung. *90*, 91-108

Zalka, K. A. 1980b The torsion analysis of an I-column with rectangular openings in both flanges. Building Research Establishment, Watford, England. N 161/80

Zalka, K. A. 1981 Combined torsional and flexural buckling of a cantilever with unsymmetric cross-section subjected to distributed normal loads. Acta Technica Hung. *92*, 189-214

Zienkiewicz, O. C., Parekh, C. J. and Teply, B. 1971 Three dimensional analysis of buildings composed of floor and wall panels. Proceedings of the Institution of Civil Engineers. *49*, 319-332

Chapter 8 (Appendices)

Bridgman, P. W.	1931	Dimensional analysis. Yale University Press, New Haven, USA
Dunkerley, S.	1894	On the whirling and vibration of shafts. Philosophical Transactions of the Royal Society of London. Ser. A *185,* 279-360
Ellis, B.	1966	Basic concepts in measurements. Cambridge University Press, Cambridge, UK
Föppl, L.	1933	Über das Ausknicken von Gittermasten, insbesondere von hohen Funktürmen. ZAMM *13*, 1-10
Hodgkinson, E.	1840	Experimental researches on the strength of pillars of cast iron and other materials. Philosophical Transactions of the Royal Society of London. 385-456
Kollár, L. Ed.	1991	Special stability problems in civil engineering (in Hungarian). Akadémia Kiadó
Merchant, W.	1954	The failure load of rigidly jointed frameworks as influenced by stability. The Structural Engineer. *32*, 185-190
Papkovics, P. F.	1963	On structural mechanics in shipbuilding. *4*, Stability of beams, trusses and plates (in Russian). Sudpromgiz, Leningrad
Rankine, W. J. M.	1863	A manual of civil engineering. 2nd edition. Charles Griffin and Comp. London
Southwell, R. V.	1922	On the free transverse vibration of a uniform circular disc clamped at its centre, and on the effects of rotation. Proceedings of the Royal Society of London. Ser. A. *101,* 133-153
Tarnai, T.	1980	Generalization of Southwell's and Dunkerley's theorems for quadratic eigenvalue problems. Acta Technica Hung. *91*, 201-221
Tarnai, T.	1981	Generalization of the Dunkerley theorem and its application in the theory of elastic stability. 9th International Congress on the Application of Mathematics in Engineering Science, Weimar.
Vlasov, V. Z.	1961	Thin-walled elastic beams. Israeli Program for Scientific Translation, Jerusalem

10

BIBLIOGRAPHY

Al-Sarraf, Z. S.	1986	Shear effect on the elastic stability of frames. The Structural Engineer. V64B(2), 43-47
Allen, H. G. and Bulson, P.	1980	Background to buckling. McGraw-Hill, London
Awadalla, E. S.	1983	Elastic critical loads of multistorey rigid frames. Proceedings of ASCE, Journal of the Structural Division. *109*, 1091-1106
Bajoria, K. M.	1986	The three dimensional progressive collapse of warehouse racking. PhD Thesis, Cambridge
Baker, J. and Heyman, J.	1969	Plastic design of frames. Fundamentals. Cambridge University Press
Barta, T. A.	1967	On the torsional-flexural buckling of thin-walled elastic bars with monosymmetric open cross section. In "Thin-walled structures," Ed. A. H. Chilver. Chatto and Windus, London, 60-86
Bazant, Z. P. and Cedolin, L.	1991	Stability of structures - elastic, inelastic, fracture and damage theories. Oxford Engineering Series, Oxford University Press
Beck, H.	1958	Ein Beitrag zur Berechnung regelmäβig gegliederter Scheiben. Ingeniour-Archiv *26*, 343-357
Beck, H.	1962	Contribution to the analysis of coupled shear walls. Journal of the American Concrete Institution *59*, 1055-1069
Beck, H. and Zilch, K.	1972	Stability in the design of tall concrete buildings. International Conference on Planning and Design of Tall Buildings. Lehigh University, Pennsylvania
Beck, H., Glück, J., König, G. et al	1978	Planning and design of tall buildings. ASCE, New York, Proceedings, CB-8, 347-381
Beck, H., König, G. & Reeh, H.	1968	Kenngröβen zur Beurteilung der Torsions-steifigkeit von Hochhäusern. Beton und Stahlbetonbau. *12*, 268-277
Bhatt, P. and To, Y. K.	1988	Assessment of the accuracy of some approximate methods of analysis of tall buildings. Proceedings of the 4th International Conference on

		Tall Buildings. Hong Kong, V1, 538-541
Billington, D. P. and Goldsmith, M. Editors	1986	Technique and aesthetics in the design of tall buildings. Fazlur Khan Memorial Session. Lehigh University, Bethlehem, Pennsylvania
Bolander, J. Jr and Wight, J. K.	1991	Finite element modelling of shear wall dominant buildings. Journal of Structural Engineering. *117*, No 6, 1719-1739
Bradford, A. M.	1988	Lateral stability of tapered beam-columns with elastic restraints. The Structural Engineer, V66 (22), 376-382
Brandt, B.	1976	Zur Beurteilung der Gebäudestabilität nach DIN 1045. Beton- und Stahlbetonbau 7/1976, 177-178
Casciati, F. and Faravelli, L.	1991	Fragility analysis of complex structural systems New York, Wiley
Chajes, A.	1974	Principles of structural stability theory. Englewood Cliff, Prentice Hall
Chan, H. C. and Cheung, Y. K.	1980	Lateral and torsional analysis of spatial wall systems using higher order elements. The Structural Engineer. Volume 58B/No. 3, September, 67-70
Chan, H. C. and Kuang, J. S.	1988	Coupled shear walls with stiff zones. Proceedings of the 4th International Conference on Tall Buildings. Hong Kong, V1, 533-537
Chan, H. C. and Kuang, J. S.	1988	Torsional behaviour of braced thin-walled open sections. Thin-Walled Structures. *6*, 343-354
Chan, H. C. and Kuang, J. S.	1989	Elastic design charts of stiffened coupled structural walls. Proceedings of ASCE, Journal of Structural Engineering. *115*, No.2, 247
Chen, W. F. and Atsuta, T	1977	Theory of beam-columns. McGraw-Hill, London
Chen, F. W. and Lui, M. E.	1987	Structural Stability, Theory and Implementation. London, Elsevier
Cheryachukin, V. V.	1968	Analysing stability of tall buildings by statistical methods (in Russian). Stroitelnaya mekhanika i raschet sooruzhenii, *11* (4), 19-21
Choudhury, J. R.	1968	Analysis of plane and spatial systems of interconnected shear walls. PhD Thesis. University of Southampton
Coull, A.	1973	Interaction between coupled shear walls and cantilevered cores in 3D regular symmetric shear wall structures. Proceedings of the Institution of Civil Engineers, 55/2, 827-840
Coull, A.	1975	Torsion of structural cores on deformable foundations. Building Science, *10*, 57-64, Pergamon Press
Coull, A.	1990	Analysis for structural design. In "Tall Buildings: 2000 and Beyond". 4th World Congress, Hong Kong
Coull, A. and Ahmed. A. K.	1978	Deflections of framed-tube structures. Proceedings of ASCE, Journal of the Structural Division ST5, 855-862
Coull, A. and Bensmail, L.	1991	Stiffened coupled shear walls. Journal of Structural Engineering. *117*, No 8, 2205-2223

Coull, A. and Bose, B	1975	Simplified analysis of frame-tube structures. Journal of the Structural Division, ST11, 2223-2240
Coull, A. and Irwin, A. W.	1971	Torsional analysis of multistorey shear wall structures. American Concrete Inst., SP 35-6, Detroit, Mich., 211-238
Coull, A. and Irwin, A. W.	1972	Model investigations of shear wall structures. Proceedings of ASCE, Journal of the Structural Division. ST 6, 1223-1237
Coull, A. and Low, K. C.	1979	Analysis of stiffened non-planar coupled shear Proceedings of ICE. *67*, Part 2, 971-986 Structural Dynamics. *2*, 171-183
Coull, A. and Tawfik, S. Y.	1981	Elasto-plastic analysis of core structures subject to torsion. Proc. ICE, *71*, Part 2, 789-804
Coull, A. and Wong, Y. C.	1981	Bending stiffness of floor slabs in cross-wall structures. Proc. ICE. *71*, Part 2, 17-35
Council on Tall Buildings	1967	Proceedings of the International Conference on Planning and Design of Tall Buildings. New York
Council on Tall Buildings	1972	Proceedings of the International Conference on Planning and Design of Tall Buildings. ASCE-IABSE, Lehigh University. Bethlehem, Pennsylvania
Council on Tall Buildings	1978	Planning and Design of Tall Buildings, a Monograph with 5 volumes. ASCE, New York
Council on Tall Buildings and Urban Habitat	1983	Development in Tall Buildings. Van Nostrand Reinhold, New York
Council on Tall Buildings and Urban Habitat	1986	Advances in Tall Buildings. Van Nostrand Reinhold, New York
Council on Tall Buildings and Urban Habitat	1986	High-rise buildings: Recent progress. Lehigh University, Bethlehem
Council on Tall Buildings and Urban Habitat	1988	Second century of the skyscraper. Van Nostrand Reinhold, New York
Council on Tall Buildings and Urban Habitat	1990	Tall Buildings: 2000 and beyond. 4th World Congress. November 5-9, Hong Kong. Editors: Beedle, L. S. and Rice, D. B.
Davies, J. M.	1979	Multi-storey plane frames. Civil Engineering, 64-69
Despeyroux, J.	1972	Analyse statique et dynamique des contraventments par consoles. Annales de l'Institut Technique du Bâtiment et des Travaux Publics. *290*
Dicke, D.	1972	Effect of fundation rotation. International Conference on Planning and Design of Tall Buildings. Lehigh University, Pennsylvania, No. 3, 61
Dowrick, D. J.	1976	Overall stability of structures. The Structural Engineer. *54*, 399-409
El-Hammasi, S. A.	1990	Numerical method for analysing open thin-walled structures under interaction of bending and torsion. Computers and structures. *37*, (6), 947-956
Faulkner, D.	1981	A convenient approach to local beam and buckling analyses. University of Glasgow. In

"Joints in Structural Steelwoork", 5.58-5.72

Fintel, M. editor 1974 Handbook of concrete engineering. Van Nostrand Reinhold Company

Gaiotti, R. and Stafford Smith, B. 1989 P-delta analysis of building structures. Proc. of ASCE, Journal of Structural Engineering. *115*, No. 4, 755-770

Ghosh, K. A., and Dilger, H. W. 1988 Interaction between the core and the subsurface periphery structure of a highrise building. Proceedings of the 4th International Conference on Tall Buildings. V1, 298. Hong Kong.

Glück, J., Goldberg, J., Wang, C., Yarimer et al 1979 Stability - in "Planning and Design of Tall Buildings". SB-4, 238-342, ASCE, New York

Goodner, B J and Gere, J M 1976 Analysis of shear cores using superelements. Proceedings of ASCE, Journal of Structural Division. *102*, (ST 1), 267-283

Goschy, B. 1971 Spatial stability of system-buildings. Acta Technica Academiae, Scientiarum Hungaricae. *70*, (3-4), 459-470

Goschy, B. 1978 Stability of core-supported structures. Acta Technica Hung. *87*, (1-2), 59-68

Goschy, B. 1979 Stability of supporting cores subjected to accidental loadings. International association for shell and spatial structures

Goshy, B. 1983 Stability control of steel structures in inelastic deformation. Stability of metal structures, Paris, Final report, 241-246

Goschy, B. 1984 Stability control of core supported structures in inelastic deformation. 3rd. International Symposium on Wall Structures, CIB, Warsaw, 133-138

Goto, Y. and Chen, W. F. 1987 On the computer-based design analysis for the flexibly jointed frames. Journal of Constructional Steel Research. *8*, 203-231

Hageman, J. G. 1967 The stability of a building. Cement. *19* (1), 4-13

Heidebrecht, C. A. and Stafford Smith, B. 1973 Approximate analysis of open-section shear walls subject to torsional loading. Proceedings of ASCE, Journal of Structural Division. *99*, (ST 5), 2355-2373

Heyman, Jacques 1971 Plastic design of frames. Applications. Cambridge University Press

Hoenderkamp, D.C.J. and Stafford Smith, B. 1987 Approximate rotation analysis for plansymmetric high-rise structures. Institution of Civil Engineers. Proceedings, Part 2: Research and Theory, V83, 755-67

Hone, C. P. 1967 Torsional-flexural buckling of axially-loaded, thin-walled, elastic struts of open cross section. In "Thin-walled structures", edited by A. H. Chilver. London

Horne, M. R. 1975 An approximate method for calculating the elastic critical loads of multistorey plane frames. The Structural Engineers. *53*, 242-248

Iffland, J. S. B.	1972	Design for stability in high rise buildings. Preprints from International Conference on Tall Buildings. V. 2-16, 39-56. Lehigh University, Pennsylvania
Jaspart, J. P.	1988	Ultimate load of frames with semi-rigid joints. Journal of Constructional Steel Research. *VII*, (4), 283-312
Jenkins, W. M. and Harrison, T.	1969	Analysis of tall buildings with shear walls under bending and torsion. Symposium on Tall Buildings, University of Southampton
Kármán, T. and Biot, M. A.	1939	Mathematical methods in engineering. McGraw-Hill, London
Kirby, A. P. and Nethercot, A. D.	1979	Design for structural stability. Constrado Monographs. St. Albans, Hertfordshire. Granada Publishing, 165
Kollbrunner, C. F. and Basler, K.	1966	Torsion. Springer-Verlag, Berlin
König, G.	1990	Hochhäuser aus Stahlbeton. In "Beton-Kalender." Wilhelm Ernst & Sohn
Kuang, J. S., Cheung, Y. K. and Chan, H. C.	1991	Elasto-plastic analysis of perforated core structures subjected to applied torque. In "Thin-walled structures," *11*, 439-456. Elsevier Sciences
Lee, L. S., Swaddiwudhipong, S. et al	1988	Effect of axial deformation on the behaviour of frame-core wall buildings. Proc. of the 4th International Conference on Tall Buildings. Hong Kong. V1, 526-532
Liauw, T. C.	1978	Torsion of multi-storey spatial core walls. Proc. ICE. *65*, Part 2, 601-609
MacGregor, J. G.	1972	Stability of reinforced concrete building frames. Preprints from International Conference on Tall Buildings. V3-23, 19-35. Lehigh University, Pennsylvania
MacGregor, J. G.	1990	Stability of reinforced concrete building frames. In "Tall Buildings: 2000 and Beyond." 1139-1148. 4th World Congress. Council on Tall Buildings, Hong Kong
MacLeod, I A	1990	Analytical modelling of structural systems. Ellis Horward
MacLeod, I. A. and Hosny, H. M.	1977	Frame analysis of shear wall cores. Proceedings of ASCE, Journal of The Structural Division, *103*, (ST 10), 2037-2047
MacLeod, I. A. and Marshall, J.	1983	Elastic stability of building structures. The Michael R. Horne Conference, 20-22 September The University of Manchester
McCallen, D. B. and Romstad, K. M.	1990	A continuum model for lattice structures with geometric and material nonlinearities. Computers and structures. *37*, (5), 795-822
Megson, T. H. G., and Harrop, J.	1975	Approximate analysis of closed section core walls containing perforations. Proceedings of the Pan-Pacific Tall Buildings Conference, 291-302, Honolulu
Mekkonen, B.	1987	Effective length and rigidity of columns.

		ACI Structural Journal, V84(4), 316-329
Melosh, R. J.	1990	Structural engineering analysis by finite elements. Englewood Cliffs, NJ: Prentice Hall
Mendis, A. P. and Darwall, LeP. P.	1987	Computer program for stability analysis of softening frames. Monash University. Civil Engineering Research. Report No.3/1987, Melbourne, The University, 35
Menzies, J.B., Armer, G. S. T. and Currie, D.	1982	The proposed new form of the Building Regulations. Building Research Establishment. Note No. 109/82
Morris, L, J. Ed.	1983	Instability and plastic collapse of steel structures. Proceedings of "The Michael Horne Conference," Granada
Nazar, K. K. and Coull, A.	1976	Elastoplastic analysis of coupled shear walls. Proceedings of ASCE, Journal of The Structural Division. ST9, 1845-1860
Peköz, T. B. and Winter, G.	1969	Torsional-flexural buckling of thin-walled sections under eccentric load. Proceedings of ASCE, Journal of the Structural Division. *95*, ST 5, 941-963
Pezeshk, S and Hjelmstad, K. D.	1991	Optimal design of planar frames based on stability criterion. Journal of Structural Engineering, *117*, No. 3, 896
Plantema, F. J.	1961	Sandwich construction. The bending and buckling of sandwich beams, plates and shells. McGraw-Hill, London
Puri, R. D.	1967	A study of the continuous connection method of analysis of coupled shear walls. PhD Thesis. University of Southampton
Qizhi, Liang and Li, Xie	1988	Strip model in analysis of tall buildings by the generalised continuum method. Proc. of the 4th International Conference on Tall Buildings, Hong Kong, V1, 309-14
Quadeer, A. and Stafford Smith, B.	1969	The bending stiffness of slabs connecting shear walls. Journal of the Americal Concrete Institute. *66*, 464-473
Rhodes, J.	1990	Thin-walled structures. Developments in theory and practice. *9*, Nos 1-4, Elsevier
Roark, R. J. and Young, W. C.	1975	Formulas for stress and strain. McGraw-Hill, London
Roberts, T. M.	1985	Section properties of thin walled bars of open cross-section. The Structural Engineer. 63B (3), 63-67
Roberts, T. M. and Achour, B.	1990	Torsion and bending of braced thin-walled open sections. Proc. ASCE, Journal of Structural Engineering. *116*, No. 1, 1-11
Rosman, R.	1969	Torsion of perforated concrete shafts. Proceedings of ASCE, Journal of Structural Division. *95*, ST 5, 991-1010
Rosman, R.	1974	Stability and dynamics of shear-wall frame structures. Building Science. *9*, 55-63
Rutenberg, A and Eisenberger, M.	1990	Stability of outrigger-braced tall buildings. In "Tall Buildings: 2000 and Beyond." 881-892.

		4th World Congress, Council on Tall Buildings, Hong Kong
Rutenberg, A., Leviathan, I. and Decalo	1988	Stability of shear-wall structures. Proceedings of ASCE, Journal of Structural Engineering. *114*, No. 3, 707-716
Salem, A. H.	1973	Stability of unbraced single-bay multistorey frames. Proceedings of ASCE, Journal of the Structural Division. *99*, (ST 2), 223-242
Schäfer, H.	1969	Die Berechnung von Hochhausern als räumlicher Verband von Scheiben, Kernen und Rahmen. Darmstadt, Dissertation
Stafford Smith, B.	1970	Recent developments in the methods of analysis for tall building structures. Civil engineering and public works review. *65* (773), 1417-22
Stafford Smith, B. and Abate, A.	1984	The effects of shear deformations on the shear centre of open section thin walled beams. Proceedings of ICE. *77*, Part 2, 57-66
Stafford Smith, B. and Gaiotti, R.	1990	Effect of precast concrete cladding on the behaviour of tall buildings. In "Tall Buildings: 2000 and Beyond." 317-332. 4th World Congress, Council on Tall Buildings, Hong Kong
Stafford Smith, B. and Lau, P. C. M.	1972	A method of assessing the static stability of panel type buildings. Proceedings of the Institution of Civil Engineers. 53/2, 77-86
Stafford Smith, B. and Nollet, M. J.	1990	The curtailed wall-frame structure: An approximate solution. In "Tall Buildings: 2000 and Beyond." 1049-1064. 4th World Congress, Council on Tall Buildings, Hong Kong
Sugijama and Kawagoe	1975	Vibration and stability of elastic columns. Journal of Sound and Vibration. *38* (3), 341-355
Tabaddor, F and Padovan, J	1991	Element nonlinearity effects on local and global stability. Computers and structures. *40*, No. 2, 401
Taylor, J. C.	1981	Semi-rigid beam connections: Effects on column design: B. 20 Code method. In "Joints in Structural Steelwork." Ed. J. H. Howlett, W. M. Jenkins and R. Stainsby, London
Thompson, J. M. T. and Hunt, G. W.	1973	A general theory of elastic stability. John Wiley, New York
Thompson, J. M. T. and Hunt, G. W.	1984	Elastic instability phenomena. John Wiley and Sons, New York
Timoshenko, S.	1945	Theory of bending, torsion and buckling of thin walled members of open cross section. Journal of the Franklin Institute. *239*, Nos. 3, 4 and 5
Timoshenko, S. P.	1953	The collected papers of Stephen P. Timoshenko. McGraw-Hill
Timoshenko, S. P. and Goodier, J. N.	1951	Theory of elasticity. McGraw-Hill
Toader, I. H. I.	1987	A more accurate evaluation of buckling loads of thin walled members in torsion and torsional bending. Thin-Walled Structures. *5*, 181-189
Trahair, N. S. and Bild, S.	1990	Elastic biaxial bending and torsion of thin-walled members. In "Thin-walled structures," *9*,

269-307. Elsevier Applied Science

Velkov, M. 1972 Ductility factor (Discussions on definitions). Proc. of the 4th European Symposium on Earthquake Engineering, London, September 5-7.

Williams, W. F. 1976 Buckling of multi-storey frames with non-uniform columns, using a pocket calculator program. Computers and Structures. *7*, 631-637

Williams, W. F. 1979 Consistent, exact, wind and stability calculation for substitute sway frames with cladding. Proceedings of the Institution of Civil Engineers. Part 2, 355-367

Wong, Y. C. and Coull, A. 1980 Interaction between floor slabs and shear walls in tall buildings. ACI, Publ. SP-63, Detroit. In "RC structures subjected to wind and earthquake forces"

Wong, Y. C. and Coull, A. 1980 Structural behaviour of floor slabs in shear wall buildings. Advances in concrete slab technology. Edited by R. K. Dhir and J. G. L. Munday

Wood, R. H. 1958 The stability of tall buildings. Proceedings of ICE, 69-101

Yeong-Bin, Y and Shyh-Rong, K. 1991 Consistant frame buckling analysis by FE method. Journal of Structural Engineering, *117,* No 4, 1053

Yuren, H. and Bozhen, C. 1989 A consistent theory for torsion of thin-walled bars. Thin-walled structures. *7*, 119-137

Zahn, J. John 1988 Combined load-stability criterion for wood beam-columns. ASCE, Journal of Structural Engineering, V114 (11), 2612-28

Zienkiewics, O. C. 1967 The finite element method in structural and continuum mechanics. McGraw-Hill, London

AUTHOR INDEX

Akesson 144
Allen 135
Ambos 113
Anderson 141
Andronov 23
Armer 22, 34, 40, 50, 83
Asztalos 41, 53
Bate 19
Beck 41, 113, 144, 145, 169
Bleich 39, 40
Bornscheuer 178
Brandt 148
BRE 16
Bridgman 36, 226
Brunel 1
BSI 28
Bubenheim 113
Chilver 3
Chwalla 40
Coull 113
Croll 104, 121
Csonka 41, 43
Danay 144, 179
de l'Orme 7, 8
Descartes 2
Dulácska 40, 50, 83
Dunkerley 184
Ellis 36, 211, 212, 226
Fattal 25
Fazakas 126, 167, 174
Föppl 184
Gellert 144, 179
Gere 39, 57, 60, 153, 167
Glück 40, 144, 179
Goldberg 41
Gordon 185
Goschy 41
Granger 1

Griffiths 4, 16, 23
Grimer 40, 83
Gáspár 104
Harman 143
Hegedüs 41, 55, 109, 133, 137, 141
Heunish 113
Hodgkinson 185
Horacek 143
Horne 39, 141
Howlett 140
Hunt 23
Irwin 113, 179
ISE 3, 19
Jones 140
Kollár, L 40, 41, 50, 83, 84, 144, 169, 183
Kollár, L. P. 41, 51, 55, 133, 137, 141, 179, 180
Komarakul-na-nakorn 3
König 113
Kókai 178
Leyendecker 25
Littler 36
Lok 141
MacLeod 113
Mainstone 7, 8, 27
Merchant 39, 185
Moore 38
Morgan 7
Nair 143
NCE 26
Papkovics 184
Parekh 143
Pearson 178
Podolski 144
Pontriagin 23
Rankine 185
Reeh 148
Renton 3
Rosman 41, 113, 122, 144, 179
Sattler 40
Schueller 113
Shäfer 144, 145, 148, 169
Sigalov 41
Southwell 183
Stafford-Smith 113
Stüssi 178
Szerémi 44
Szittner 104, 121
Szmodits 144
Taranath 113, 144, 179
Tarnai 141, 183
Teply 143
Thom 33
Thompson 23

Timoshenko	39, 57, 60, 153, 167
Tornyos	104
Vlasov	179, 192
Walker	104, 121, 143
Wang	40
Weaver	39
Wood	36, 40, 141
Yarimci	144
Zalka	46, 51, 52, 56, 59, 62, 74, 84, 115, 124, 129, 130, 148, 157, 179
Zienkiewitz	39, 143

SUBJECT INDEX

abnormal load 31, 35, 41
accidental actions 4
accidental load 16, 19, 30, 31
accuracy 39, 40, 41, 62, 76, 80, 85, 104, 106, 112, 117, 121, 125, 128, 141, 169, 184, 219
accuracy analysis 44, 67, 83-84, 92, 97, 117, 123, 141, 161
additive formula 55, 72, 76, 84, 85, 87, 88, 92, 93, 106, 107, 116, 117, 133, 166
additive theorem 49, 60, 66, 71, 73, 80, 85, 124, 144, 154, 155, 163, 168, 183
alarm bells 38
alternative path 15, 28, 29, 30
analogy 64, 128, 130, 132, 135, 137, 139
analytical method 2
analytical models 34
analytical theories 33
anistropic construction 19
applied loads 81, 88, 111, 126, 128, 143, 163, 168, 175
approximate analysis 132, 178
approximate critical load 66, 85, 88, 89, 118-123, 132, 144
approximate formula 55, 62, 66, 84, 89, 94, 100, 118, 126, 132, 163, 184
approximate method 40, 92, 95, 102, 113, 157, 179
approximate procedures 40, 84, 90, 96, 99, 106, 128, 144
approximate solution 62, 74, 80, 124, 128, 130, 161, 185
approximation 40, 44, 50, 52, 59, 66-68, 76-77, 81, 111, 112, 119, 141, 157, 161, 166, 173, 176, 178, 180
arbitrary distribution 133
arbitrary shape 62, 113, 150
arrangement of structural elements 5, 7, 8, 10, 11, 13, 13, 15, 143, 144, 147, 148, 152, 169, 173, 178
assumptions 40, 41, 68, 84, 114, 144, 178
asymmetric 41
axial deformation 41, 43-44, 52-57, 66, 71, 79, 80, 147, 179
axial forces 44, 79
bays 39, 41, 44, 52-55, 82, 84-85, 89, 94, 97, 99, 104, 106, 109, 113

beam stiffness	84, 109, 114, 115, 116, 119, 138, 139, 140
beam-column joints	109, 139, 140
behaviour of structure	5, 7, 13, 14, 15, 21, 29, 31, 34
bending	51-53, 92, 114, 119-121, 147, 151, 152, 156, 178, 179, 189, 192
bending analysis	176
bending buckling	89
bending deformation	48, 49, 53, 61, 62, 125, 128, 129-131
bending resistance	66, 132
bending shape	47
bending stiffness	40, 42, 60, 61, 63, 70, 73-75, 78-79, 129, 131-138, 144, 148, 176, 179, 181
bending stiffness of the beams	41, 46, 47, 49, 60, 70, 106, 114
bending stiffness of the columns	40, 42, 49, 61, 70, 78
bending type	84
Bessel functions	60, 153, 157
bifurcation-point	39, 41
boundary condition	47, 48, 56, 59-61, 69-74, 113, 129-131, 151, 156, 187, 192, 193, 210
boundary-value problem	62
British codes	28
brittle	11, 13, 14, 19, 36
BS	28
buckling	41-47, 50, 52, 60, 65, 66, 70-71, 81, 106, 111, 124, 143, 148, 150, 156, 162, 189
buckling mode	177
buckling shape	81, 84, 89, 91-95, 98-103, 111, 118-121, 124, 184
Budapest Technical University	104, 121
Building Research Establishment	37
Camden school roof	16
cantilever	42, 52, 55, 57, 60, 62-64, 92, 128-130, 135, 143, 163, 168, 188
cantilever, substitute	59, 60, 144, 146, 149, 179
centrally compressed structure	152
centre of load	59, 112, 149, 156, 175
centre of support system	124, 176, 192
centre of torsion	178
centroid	150, 156, 163, 170, 175, 192
centroidal axis	53, 54, 114, 135, 150, 174, 185
change in length	49, 53
characteristic deformation	48, 65, 84, 98, 123, 125, 131
characteristic mode	84, 177
characteristic shape	95
characteristic (stiffness) ratio	43, 89, 92, 94-97, 99, 100
cladding	141
clamped	144, 192
classical method	36
classical theory	107
classical tools	2
closed cross-section	144, 156, 176, 178, 180, 181
closed form solution	39, 41, 53, 60, 70, 73, 133, 143, 157
closed form formula	45, 48, 114

code of practice 5, 19, 27, 40, 88, 128, 140, 173
collapse 4, 5, 8, 19, 24, 25, 33, 37
collapse load, rigid-plastic 140, 141, 185
column lengthening 44, 52, 53, 66, 71, 79
column shortening 44, 52, 53, 66, 71, 79
compatibility conditions 39
component structures 62, 183
compressive forces 40, 152, 188
compressive stresses 188, 192
computer procedure/method 39, 104, 106, 121, 126, 167, 174
concentrated force 39, 41-45, 53, 57-59, 62-69, 76-99, 104, 106-107, 110-111, 117, 121, 124, 126-128, 133, 136, 139
concentrated load 48, 64, 80, 85, 86, 88, 89, 111, 117, 118, 126, 167, 175
concentrated moments 46
conditions 43, 44, 45, 61, 115, 159
connecting media 135, 136
connectivity 15, 35
conservative critical load 62, 63, 80, 84, 144
conservative estimate 67, 81-83, 85, 105-106, 118, 123, 157, 179
conservative forces 41
conservative formula 76, 133, 141, 155, 163, 184
continuity 45, 224
continuous 46, 58, 92
continuously distributed 50, 64, 70
continuum method 39, 109, 123, 180
continuum model 42, 45, 50-52, 56-59, 64, 67-73, 106, 141, 149
continuum solution 64, 77, 121, 133, 155
contraflexure point 42, 44-46, 52, 53, 58, 66, 70, 114, 115
control systems 19, 38
convergence 157
core 126, 143-149, 156, 169, 173-181, 192
correction 81, 82, 111, 168
Council on Tall Buildings 40
coupled shear walls 41, 109, 113-126, 141, 143, 178-180
criteria 68
critical load 48, 53, 55, 57, 58, 66, 67, 72, 80-81, 87, 88, 104-106, 110, 111, 116-126, 132-136, 140, 144, 153, 156-157, 162, 165, 167-168, 180, 183-185, 192
cross wall building 109, 124, 125, 126, 127
cross-sectional area 53, 114, 129
crosswalls 29, 30
damage 5, 19, 24, 25, 26, 28, 29, 30, 31
de l'Orme 7, 8
dead weight 167
debris loading 15, 24, 26
deflection, sudden 62, 74
deflection curve 46, 61, 70, 114, 115, 131
deformation 49, 50, 51, 53, 61, 62, 65, 79, 84, 104, 106, 114, 123, 143, 184

deformation characteristics 13, 14, 19, 20, 22, 36
design 33, 17, 19, 27, 30, 31, 35-38
design codes 17, 28
design example 84, 117, 126, 169, 180
design method 2, 27
design office 40, 209
design process 2, 7, 34, 35, 37, 86, 88, 163
design rules 5, 27, 28, 31
discontinuity 209, 217, 223, 226
discrepancy 55, 84
displacement 185, 187
distributed bending stiffnesses 46, 47, 49, 61, 70, 114, 116
distributed, downwards 52, 59
distributed moments 41-42, 46, 69
domestic gas 17, 31, 38
dominant 67
dominant global bending 93, 106
dominant shear 89, 106
dominant warping stiffness 180
door openings 119, 176
double curvature bending 44, 45, 50, 52, 84
doubly symmetric 152, 163, 165, 177, 180
ductility 11, 13, 14, 15, 17, 20, 23
Dunkerley correction 81, 82, 111
Dunkerley theorem 80-81, 87, 106, 111, 117, 125, 140-141, 168, 173, 184
dynamic analysis 178
dynamic problems 41
earthquake 14, 22, 31, 179
economic structure 40, 126, 157
effective length 40, 51, 70, 83, 110
efficiency 143
eigenfunction 62
eigenvalue 48, 49, 60-62, 67, 70, 74, 124, 130, 144, 153, 157, 160-161, 165, 167, 184, 192
elastic 40, 139-141, 144, 183-185
elasto-plastic 139, 140, 141, 185
electronic computation 40
elementary section 46, 47, 49, 56, 129, 150, 186, 188
elementary statics 78, 96, 102, 187, 189
elements of law 209
end-conditions 40
energy method 39
equilibrium 44, 46, 49, 51, 52, 56, 69, 78, 79, 96, 102, 129, 150, 186, 188
error 55, 59, 64, 77, 80, 83-85, 89-108, 112, 118-123, 134, 139, 155, 161, 178, 184, 214
essential elements 3
Euler 5, 7, 167, 168, 170, 179
Eurocodes 34
expansion joints 5, 19, 36
experiments 7

explosion 14, 15, 16, 19, 27, 29, 31, 38
faces 55, 135, 136, 137
failure 65, 71, 84, 89, 95-96, 106, 110, 124, 163
FE procedure 89, 96, 102, 106, 108, 118, 123, 128
feedback of information 17, 19, 20
fictitious equilibrium equation 52
fictitious load 189, 190, 192
finite difference method 62, 157
finite element method 84, 107, 108, 113, 128, 143, 157
fire 31
first storey columns 111
fixed support 42-45, 64, 70, 71, 74, 79, 81, 84-85, 106, 111, 117, 126, 137, 150, 168, 185
fixing 26, 28
fixing effect 42, 44, 46, 50, 52, 59, 69, 70, 92, 140
flexural bending 166
flexural buckling 144, 146, 152, 157, 163, 166, 170, 172, 174, 176, 180, 187
flexural critical load 180
flexural-torsional 161
floor level 44, 58, 69, 72, 110, 116, 148-9, 168, 178
floor slabs 124, 128, 143, 144, 147, 148
Föppl-Papkovics formula 55, 72, 84, 88, 107, 116, 117, 133, 134, 161, 163, 166
Föppl-Papkovics theorem 51, 66, 71, 80, 118, 130-1, 144, 157, 184
four-bay 138
Fourier series 157
frameworks 39, 44, 68, 85, 92, 104-106, 109-111, 124, 126, 128, 135, 141, 143, 153, 157, 168, 180, 185
fusing mechanisms 36
gas 27, 31, 38
generalization 41, 57, 84, 109, 141
generalized bar models 141
generalized Dunkerley formula 141
generalized method 157
geometrical arrangement 148
geometrical characteristics 104, 109, 113, 126, 140, 146, 161, 174, 178, 180, 192
geometrically perfect structure 144
global behaviour 40
global bending buckling 66, 89, 135
global bending critical load 54, 58, 72, 79, 86, 138
global bending deformation 89, 93, 95, 98, 106, 123
global bending stiffness 53, 55, 135, 136, 138, 139
global bending sway 54, 57, 79
global inertia 54
global second moment of area 66, 79, 87
global shear 49, 89, 93, 115
global shear critical load 62
global shear deformation 89, 93, 95, 98, 106, 119-123, 141
global shear effect 65
global shear stiffness 57, 106, 153

global stability analysis 180
global sway 44, 53-55, 57, 70
global translation 50
gravity load 60
ground floor beams 68, 92, 94, 97, 106, 110, 111
ground floor columns 106, 107, 111
ground floor level 59, 61, 68, 72, 76, 95, 99, 100, 102, 111, 137, 178
ground plan 124, 145, 148, 151, 169, 173, 180, 192
hierarchical structure 215
homogeneous 41, 60, 69, 144, 160
Hooke's law 41, 209, 210
horizontal load 41, 113, 141, 143
horizontal translation 47, 129
identical deflection 46, 147, 179
identical critical load 179
ill-conditioned stiffness matrix 108, 124, 157
impact 16, 31, 38
imperfections 41
in-plane buckling 63, 88, 124
incorrect load factors 124
individual cantilevers 52
individual columns 40, 48, 49, 92, 107, 185
individual elements 40, 84, 125, 146, 147
individual frameworks 126
infill panels 17, 37
infinite series 164
infinitely flexible 124, 147, 148
infinitely stiff 50, 64, 70, 71, 79, 84, 110, 124, 128, 130, 147
initial stresses 113
initial-value (problem) 62, 73
input data 39, 107, 113, 128, 143
interaction 143, 180
interpolation 165, 172
inverse addition 183, 184
isotropic 41, 144
iterative analyses 2
joints 5, 15, 19, 27, 35, 36, 41, 109, 139, 140
key elements 31, 32
lateral instability of a storey 40
lateral stiffness 113, 143
lateral translation 50, 59
law 209-226
law classes 217-223
layers of sandwich column 135
layers of top storey level 80
layout 124, 125, 127, 152, 156, 163, 174, 176
lengthening 44, 52, 53, 66, 71, 79
limitation (of method) 39, 112, 113, 183
lintels 114, 118, 143
load, conservative 41
load bearing elements 143, 144, 169

load combinations 8, 23
load movement 15
load parameter 62-64, 75-77, 87, 117, 133-135, 153-155, 161, 162, 165, 184, 192
load paths 15, 17, 20, 29, 35
load redistribution 16, 83, 103
load transmission 15, 16
load-carrying mechanism 5, 14, 15, 17, 19, 20, 31
load-deformation 13, 19, 20, 21, 22, 36, 37, 221
load-environment 21
load-shedding elements 19
load-softening 20
load-stiffening 20
loading 80, 83, 102, 107, 116, 119
loading machine 21
local bending 65, 84
local bending critical load 49, 62, 79, 86, 123
local bending deformation 92, 106, 123, 141
local bending stiffness 70, 135, 137, 153
local buckling 57, 68
local co-ordinate system 146
local critical load 55, 57, 70, 71
local damage 5, 24, 25, 26, 31
local failure 31, 111
local shear buckling 111
local shear critical load 50
local shear deformation 91, 95, 97-100, 106, 107, 123
local shear stiffness 55, 110, 111
local sway 44, 49-54, 57, 64-65, 70-71, 76-78, 95, 98, 110, 115
longitudinal fibres 53, 66, 71, 147, 148, 192
longitudinal strip 187, 188, 190
low-rise 81, 98
lower bound 60, 62, 67, 73, 74, 124, 131, 144, 157 161, 179, 183
maintenance 17, 19
material 41, 113, 144, 148
mathematical models 3, 33, 209, 215
mathematical modelling 209
matrix form solution 39, 160
matrix, ill-conditioned 108, 124
maximum deviation 63, 76, 133, 155
maximum error 80
measurement 211-215, 217, 220-223, 225
mechanism 5, 14-20, 31, 34, 36, 37, 215, 221, 226
medium-rise 55
member stiffness 43
Merchant-Rankine formula 140, 141, 185
minimum strain energy 143
model, perspex 104, 105, 121, 122, 123
modifications 114, 116
modified second moment of area 115
modified shear formula 115, 116

modified beam stiffness	140
modifier	67, 71, 72, 79
modulus of elasticity	45, 53, 84, 118, 127, 129, 147, 170, 173
modulus of elasticity in shear	114, 118, 127, 129, 147, 170, 173, 190
moments, uniformly distributed	41, 42, 46, 69
monitoring	19, 38
monosymmetric	152, 156, 161, 162, 164, 165
multi-bay frameworks	39, 52, 53, 54, 135, 137
multi-layer	55, 135, 137
multiple mechanisms	14, 17, 19, 35
multistorey	39, 41, 53, 80, 168
negative error	85, 118
nodal points	84
non-linear	39, 40
non-loadbearing elements	17
non-proportional framework	85, 89, 92, 97, 104, 109, 110
non-proportional loading	83, 102, 103, 107
non-regular	97, 107, 109, 112
non-uniform distribution of load	67
non-uniform torsion	190
normal forces	46, 66
normal loading	39, 143
normal stresses	41
numerical analysis	154, 55, 84, 106, 169
numerical example	81, 111, 117, 138, 169, 170, 174, 176
numerical method	62
numerical proof	211, 212
numerical problems	60, 108, 124
numerical solution	74
one-bay framework	41, 44, 53, 55, 104, 135
one-bay shear wall system	121, 122
open cross-sections	144, 152, 176, 178, 179, 180
open domain	61
openings	124, 176, 178
output data	39
overall critical load	81, 88, 106, 123, 126, 144, 173-177, 180-1
overall stability analysis	144, 168
overall warping stiffness	148
overestimated critical load	59, 102, 106, 107, 108, 123, 168
overestimation	55, 59, 102
overloaded columns	83, 102, 103, 107
overloading	19, 83, 102, 126
overturning	5
pack-of-cards	15
panel structure	27, 37
panorama restaurant	167
partial collapse	4
partial factors	34
partial stiffening	184
perspex model	104, 105, 121, 123
pierced support elements	118, 178, 179
pin-ended columns	73, 169
pinned bars	147

pinned columns 173
pinned joint 140
pinned support 43, 68-80, 92, 96-98, 101-102, 106, 108, 110, 128, 132, 137, 185
planar buckling 126, 152
planar stability analysis 124
plastic hinges 15
plot 104, 121
point of contraflexure 42, 44-46, 52, 53, 58, 66, 70, 114, 115
polar moment of inertia 151, 190
positive error 85, 118, 123
power series 74, 133, 157, 158, 159, 164
practical application 40, 43-44, 80, 84, 102, 111, 139, 156-157, 162, 165, 168, 178, 180
principal axes 150, 152, 172, 176, 185
principal planes 124, 143, 163, 172, 175, 180
problem of dimensions 215
progressive collapse 5, 24, 37
progressive failure 25, 26, 27, 38
proportional 43, 44, 53, 58, 69, 84, 92, 94, 109, 111, 112, 120, 121, 138
proportional load 44, 69, 83, 102, 103, 94
proportionality 44, 46, 68, 84
proposition 15, 17, 22
protected elements 31
pure torsional buckling 63, 124, 143, 148, 152-153, 155, 157, 162, 163, 165, 168, 180, 191
rack structures 144
radius of gyration 148-149, 154, 158, 171, 175, 180, 191-192
Rankine formula 140, 141, 185
rapid check 40, 126, 137
re-distribution 83, 103
reciprocal addition 51, 131
reciprocal formula 55, 72, 84, 111, 116, 125, 184
reciprocal theorem 66, 71, 80, 131, 144
rectangular cross-sections 114, 129
rectangular ground plan 148, 192
rectangular network 41, 44, 68
recursion 165
redundancy 15, 17, 35
regular coupled shear walls 113
regular framework 39, 41, 44, 68, 80
regular load 39
regular structure 143, 144, 178
regularity 44, 80, 144
reserve capacity 20
right-end beam stiffness 115
right-hand rule 191
rigid joints 41, 50
rigid-body motion 102
rigid-plastic 140, 141, 185
robustness 4, 5, 19, 23, 31, 34, 36, 37, 38
Ronan Point 4, 16, 23, 27, 29, 30

roof level 167
rotation 50, 150, 151, 156, 181, 185
rough estimates 137
rule classes 30, 31, 215, 221, 223, 225
safe approximation 76, 81, 106, 163
safety 31, 33, 34, 81, 88, 126, 128, 173
sandwich analogy 139
sandwich column 41, 55, 109, 128, 135, 136, 137
scale of measurement 211, 215
scientific method 3, 7
second moment of area 45, 47, 53-59, 66, 70-71, 79, 87, 110, 115, 120, 129, 135, 138, 146, 190
secondary effect 57, 66, 70, 72
semi-rigid joints 109, 139, 140
serviceability 33
shear buckling 62
shear centre 145-153, 156, 163-166, 170, 178, 181, 185, 186, 187, 190, 192
shear critical load 49, 50, 51, 52, 62
shear deformation 48-51, 61-62, 70-71, 84, 89, 91, 93, 95, 97-100, 106-107, 114-115, 125, 128-129, 178, 179
shear force 42, 44-48, 52-53, 60, 66, 70, 72, 79, 130, 143, 151, 187
shear resistance 61, 65, 132
shear stiffness 49-52, 55, 57, 58, 60-65, 70, 72-79, 86, 106, 114, 116, 132-134
shear stiffness, modified 116
shear sway 57, 89, 93, 95
shear walls 41, 113, 118, 120, 126, 143-144, 148, 169, 179, 180
shortening 44, 52, 53, 66, 71, 79
sign convention 47, 191
simplifying assumptions 40, 144
single cantilever 59, 144, 146
single column 40, 42, 46
single core 144, 156, 169, 173, 181
single-bay structures 113
singly-symmetric 156
Skyline Plaza 25
slim frameworks 54
small cross-section 57
small deformation 41, 144
small-scale test 84, 104, 118, 121
softening 53
Southwell's plot 104, 121
space frames 41
span-to-depth ratios 114
spatial analysis 143, 179, 190, 192
special sums 146
St Venant constant 147
stability 17, 29
stable 175, 177

stable procedure 62
state of stress 40, 178, 179
statically applied load 41
statically determinate structure 41
statically indeterminate structure 41
step-by-step method 40
Stepney swimming pool 18, 19
stiffness 41, 43, 183, 184
stiffness, infinitely great 50, 64, 70-71, 78-79, 84, 110, 144, 147, 184
stiffness, zero 49, 60, 61, 73, 74, 75, 144
stiffness characteristics 60, 67, 73, 80, 84, 85, 90, 95, 108, 144, 149, 157, 161, 167, 170, 174, 179
stiffness regions 60, 67, 79
storey region 57, 64, 76, 81, 98, 110, 111
strain-hardening 141
strain-softening 20
strain-stiffening 20
strategies for design 17, 19, 27, 30, 31, 35, 36, 37, 38
stress analysis 41, 113, 125, 143, 169, 187, 189
stress-strain curve 223, 225
stronger members 40, 83
structural design 1, 2, 7, 33, 35, 85, 88, 126, 141, 173, 181, 209
structural engineer 3, 15, 23, 31, 36, 40, 41, 112, 113, 126
structural form 1, 2, 3, 5, 142
structural integrity 4, 5, 28, 36
structural properties 15, 34
structural systems 7, 19, 23, 34, 35
structure-environment 8, 15, 17, 34
sub-structures 143
substitute cantilever 59, 60, 144, 146, 149, 150, 179
substitute column 41-42, 44-52, 56, 59-60, 62-75, 78, 111, 141, 144, 147, 150, 167, 178
substitute continuum 50, 67
substitute core 145
substitute load 59, 72, 112
substitute thickness 179
successive approximation 40, 161
successive determination 159, 165
superimposed parts 183
support conditions 4, 34
support elements 124, 127, 146, 148, 178, 180, 181
support system 124, 126, 135, 144-147, 152, 157, 169-170, 175-176, 178, 180-181, 192
sway structures 41, 144
swimming pool 167
symmetric 41, 44, 55, 122, 166
symmetric arrangement 124, 126, 143
symmetry 166
synthetic process 2
system theory 35
tall buildings 41, 144, 178, 179

tall frameworks 54
tall structures 81
tertiary effect 40, 41
test 84, 104, 105 118, 121
test data 209
thick faces 55, 135, 137
thin faces 135, 137
thin walled cross-section 185, 192
thin-walled closed cross-section 156
thin-walled open cross-section 152, 179
thin-walled structures 143, 144, 168
tie forces 29, 30
ties 5, 28, 29, 30, 37
torque 148, 179, 190, 191
torsion 148, 151, 156, 178, 187, 192
torsional buckling 64, 126, 143-149, 157, 163, 179, 185
torsional constant 147, 190
torsional deformation 178
torsional moment 152
torsional stiffness 64, 144, 146, 148, 153-155, 171, 176-177, 179-181, 191-192
torsional-flexural 152, 157, 161, 162, 166, 168, 187
total critical load 48, 83, 124
total load 40, 42, 59, 76, 88, 126, 170, 173
translation 45, 47, 50, 53, 59, 114-115, 129, 150, 185, 186, 188
translation, sudden 131
transverse connections 16, 19
twist 156, 190, 192
two-bay framework 82, 85, 97
two-bay symmetric frameworks 41, 44, 55
tying 19, 28, 29, 30, 31
U-shaped core 174
UDL 58-59, 65-67, 72-73, 80-99, 102-107, 111, 116-117, 126, 129, 132, 136, 138, 151, 167-171, 173-176
ultimate load 33
underestimated critical load 106, 107, 123
underloaded bars 83, 103
uniformly distributed load 39, 58, 80-82, 110, 116, 124, 129, 149, 150
unsafe 106, 108
unstable 175, 181
unsymmetric 152, 163, 166, 180, 185
variable cross-section 45, 55, 56, 57, 59, 60, 73
vehicle impact 16
venting 19, 27, 28, 38
vertical equilibrium equation 44, 52
vertical load 40, 47, 59, 83, 95, 106, 111, 124, 129, 141-144, 148, 152, 156, 163, 168, 192
vibration problems 183, 184
Vierendeel girders 41
wall thickness 113, 174, 175, 176

warping 148, 151, 179, 192
warping characteristics 144
warping constant 191
warping deformation 147
warping stiffness 64, 147-148, 153, 171-172, 176-177, 179, 180-181, 191-192

weak beams 101
weak columns 40, 83, 91, 95, 100
weight 56, 167, 168
whole storey 83
wind load 41
window openings 118, 176
Young's modulus 129, 216